Risk Management in Post-Trust Societies

EARTHSCAN RISK IN SOCIETY SERIES
Series editor: Ragnar E. Löfstedt

Risk Management in Post-Trust Societies

Ragnar E. Löfstedt

publishing for a sustainable future

London • Sterling, VA

Hardback edition published by Palgrave Macmillan in 2005
Paperback edition published by Earthscan in the UK and USA in 2009

ISBN 978-1-84407-702-1

Printed and bound in the UK by TJ International, Padstow
Cover design by Yvonne Booth

For a full list of publications please contact:

Earthscan
Dunstan House
14a St Cross St
London EC1N 8XA
UK
Tel: +44 (0)20 7841 1930
Fax: +44 (0)20 7242 1474
Email: earthinfo@earthscan.co.uk
Web: **www.earthscan.co.uk**

22883 Quicksilver Drive, Sterling, VA 20166-2012, USA

Earthscan publishes in association with the International Institute for
Environment and Development

A catalogue record for this book is available from the British Library

Library of Congress Cataloging-in-Publication Data has been applied for

The paper used for this book is FSC-certified and
totally chlorine-free. FSC (the Forest Stewardship
Council) is an international network to promote
responsible management of the world's forests.

FSC
Mixed Sources
Product group from well-managed
forests and other controlled sources

Cert no. SGS-COC-2482
www.fsc.org
© 1996 Forest Stewardship Council

To Laura, Anneli and Åsa

Praise for Ragnar E. Löfstedt, *Risk Management in Post-Trust Societies*

'This is a thought-provoking and invaluable book for anyone who cares about risk communication and management in the 21st century. Professor Löfstedt, via a number of case studies and the latest theoretical analysis, offers new insights on how regulators and policy-makers can best win back the public's trust in the era of post trust.'—**Anna Jung**, *Director General, European Food Information Council*

'Professor Ragnar Löfstedt has once again produced a most interesting book on risk management and trust, well-based in theory and build on empirical findings. Löfstedt presents a number of important and thought-provoking conclusions on how to integrate trust into risk management, conclusions that once implemented might not only contribute to restored trust in society, but also to decreased risk.'— **Mikael Karlsson**, *President, Swedish Society for Nature Conservation*

'Ragnar Löfstedt's book is about the need to re-establish public trust in policy-making for our modern societies. Understanding and communicating risk is essential for politicians and regulators in Europe and worldwide. In an increasingly diverse and interrelated world, a sophisticated risk management system will be the essential tool to re-establish trust. Professor Löfstedt's deep and profound scientific knowledge and his political insight into the various cases will help us to develop more appropriate risk management strategies.'— **Erika Mann**, *MEP, Germany*

Contents

List of Figures

Foreword

Human evolution has armed us with a repertoire of emotion-based abilities to process information about others rapidly and unconsciously, detect if they are attempting to deceive us, and determine if they are trustworthy. For the most part these abilities provide us with adequate guidelines for conducting the business of everyday social life. By and large, the majority of us, most of the time, are able to conduct our social lives with family, friends and acquaintances relatively smoothly by sorting out the trustworthy good from the distrust-deserving bad. The management of societal risk controversies moves us from face-to-face encounters in small groups, the environment of the evolutionary adaptation of trust, to a different arena of interaction. In this often impersonal arena of conflicting ideologically based group perspectives, common everyday understandings about trust, the folklore of trust, if you will, often let us down. It is in conflicts about environmental hazards such as those analysed in this book that we most need to shift to a different venue of understanding. Successful operation in this arena requires that implicit everyday common understandings of trust be generously supplemented with explicit uncommon understandings consisting of empirically supported systematic conceptualizations. This is especially so when valid understandings are most needed to direct relationships with the public and develop organizational responses.

Risk Management in Post-Trust Societies joins the increasing number of efforts adding to the systematic empirical-based understanding of trust. The effort here focuses on case studies in four countries including hydro-dam relicensing, siting of a waste incinerator, management of a nuclear power plant, and disposal of a North Sea oil storage buoy. Ragnar Löfstedt has been personally involved in research with each of the cases. The basic orientation of the book, reflected in its title, represents a relatively new view on trust and risk management. The eagerness to provide immediate simple answers to practical questions has spawned its own folklore about trust. A prominent current piece of this folklore goes beyond the basic recognition that trust engenders efficient management through the production of social capital. It is concluded that management is not possible without trust. Often this notion works itself out in suggestions of vapid populism where efforts to increase citizen involvement take precedent over effective risk management. One example to the contrary presented in the book is the Swedish Chemical Inspectorate implementation of strict regulations of the use of anti-fouling paints on pleasure boats. The view that efforts should focus on understanding how to manage when there is no trust or there is distrust is still somewhat contrarian. Certainly it is contrary to the accepted folk lore that management can be improved through the application of a few simple things that are presumed to increase or recover trust.

If we are to pursue seriously an understating of risk management in post-trust societies, what guidelines should we followed? Ragnar Löfstedt offers us a few valuable examples in this regard. Obviously our efforts should be informed by conceptualizations of government and management approaches. This would include an awareness of the values inherent in different philosophies of government as well as a consideration of the advantages and disadvantages of various management approaches. *Risk Management in Post-Trust Societies* presents comparisons of management approaches including deliberations involving public participation, technical risk management and economic-based cost/benefit approaches as well as discussions of national differences in risk management.

Our efforts to understand post-trust risk management should explicitly define concepts and empirically test those conceptualizations. Another part of current folklore is that there are a very large number of conceptualizations and definitions of trust (which is true) and that this confusion is indeterminate (which is not true). The existing babble will remain indeterminate and new putative distinctions will continue to be creatively spawned unless conceptualizations are grounded in broader understandings that are empirically tested. In *Risk Management in Post-Trust Societies*, for example, both 'trust' and 'risk management factors' are explicitly defined. Both are tested against the evidence of the case studies. This analytic effort yields evidence that the particular values that affect trust may vary in different circumstances. In some cases, impartiality of risk managers is extremely important. In other cases, it is not, but competency or another value might be.

Risk Management in Post-Trust Societies offers a good example of two other important guiding principles: 'pragmatism' and 'provisionalism'.

While rejecting overly-simple quick answers to practical problems, a series of methods that offer a good chance of success for risk managers are developed. These are offered with recognition that current understandings are always provisional and highly likely to be revised or replaced as our efforts for better understanding continue.

Finally, as suggested in *Risk Management in Post-Trust Societies*, we should operate on the understanding that, while important, trust is not the whole story of successful risk management. Public reliance is a necessary component of effective risk management in democratic societies. Effective management requires that the public rely on technical risk assessments and formulated risk mitigation plans. Trust results from a perception of shared values and people are likely to rely on risk managers who are trusted. People are also likely to rely on risk managers, even ones who are not necessarily personally trusted, who are perceived to be bound by institutional structures operating in the public's interest. Institutional structures, such as systematically enforced laws, procedures attempting to ensure fair and just decisions, institutionalized accountability, and effective opportunities to voice one's view provide assurances that risk assessments and mitigation plans can be relied on. Applying the arguments of practical government theorists such as David Hume and James Monroe leads to concluding that institutional structures should be designed to eliminate the need for trust or at least reduce its risks even if the risk managers are knaves.*

GEORGE CVETKOVICH
Department of Psychology
Western Washington University

---

*Hardin, R. (2004). *Distrust.*, 2004, from http://www.russellsage.org/Merchant2/working_papers/Distrust.doc

Preface to the Paperback Edition

Following the publication of the hardback edition of this book in 2005, trust and the lack of it has remained high on the agendas for regulators, policy-makers and industrialists on both sides of the Atlantic. Although the issues and arguments around trust that were raised in this book were more or less accepted by academics, it is clear that the discussion has moved on somewhat. In this brief introduction, I summarize three new trends within the area of risk trust research and in conclusion provide a few predictions on where I see the trust field moving to over the next few years.

Three trust trends

Not having trust in society may actually not be all 'bad'. Since 2005 one of the more interesting research avenues is that of the 'critical trust' (Pidgeon et al, 2007). As Pidgeon et al note:

> Critical trust lies on a continuum between outright scepticism (rejection) and uncritical emotional acceptance. Such a concept attempts to reconcile the actual resilience by the public on institutions, while simultaneously possessing a critical attitude towards the effectiveness, motivations, or independence of the agency in question. (Pidgeon et al, 2007, p127)

Pidgeon et al tested this concept via a number of large quantitative surveys in the UK and found the concept validated (e.g. Poortinga and Pidgeon, 2003a, b). In a sense, these findings indicate that distrust is not in itself necessarily undesired (for an excellent discussion on this point see O'Neill, 2002). Critical distrust can in effect be healthy for society, as the public may have become more knowledgeable and competent over time and hence be able to make up their own minds (Barber, 1983). The question remains, however, if this is always and everywhere the case, or if the situation is unique to the UK, a country that has been plagued by regulatory scandals (and hence distrust) for a long time (UK Strategy Unit, 2002; Lofstedt, 2004). Arguably, in countries such as Sweden, where there has been much greater trust in regulators for many years, critical trust would be seen by the establishment as unhealthy and worrying.

One issue that is much debated within policy and academic circles is that of increasing transparency in the policy-making process. Research findings do indicate that openness (transparency) and honesty increase trust while secrecy destroys it (Peters et al, 1997). Schütz and Wiedemann (2000), for example, demonstrated that chemical companies that offered more information about their operations were more trusted than those that did not. Similarly, Matsumoto et al (2005) found that corporations providing balanced information on nuclear power, discussing both its risks and benefits, were more trusted than those providing only positive communications. Although there is no doubt that transparency and openness promotes trust, the implementation of transparency programmes is not unproblematic. Transparency initiatives encourage members of the public, both directly and indirectly, to make their own decisions about risks; for example, deciding what food to eat rather than relying on science-based regulators to decide for them. Greater transparency also leads to the development of policy vacuums that are quickly filled by the most efficient communicators, which are not always the regulators (Lofstedt, 2004). Even more worryingly, complete transparency, such as putting out unfiltered scientific findings on the web, can lead to public alarm, with drastic public health consequences, as witnessed in the pharmaceutical sector.

In 'post-trust societies' public trust does not simply disappear altogether, but is rather re-allocated. Previously, publics trusted regulators and industry; today, the same publics trust individuals or special interest groups, who are perceived not to have vested interests in the question at hand. In the UK, for example, when it comes to environmental/energy issues such as global warming or nuclear power, the general public trusts the views and expertise of non-governmental organizations (NGOs) such as Greenpeace and independent academics more than scientists working for a utility company or for the regulator (House of Lords, 2000). As a result, these NGOs and academics have become rather powerful in the setting of regulatory agendas. For example, one so-called meta-analysis of the type 2 diabetes drug Avandia, published by highly trusted academics (see Nissen and Wolski, 2007), led to an overnight share price fall of 8 per cent for the manufacturer of that drug, the pharma company GlaxoSmithKlein (GSK). In addition, the front page media headlines reporting that the meta-analysis of Avandia showed a supposed 43 per cent increase in heart attacks led to approximately half a million Americans stopping taking that type of diabetes medicine. In the resulting media and policy-making discussions, the primary focus was on Dr Nissen rather than on the pharma regulators and industry (who had conducted their own independent meta-analysis; Lofstedt, 2008). Had such a study been published only ten years ago, the main regulator – the US Food and Drug Administration, at the time highly trusted by the US public – would have quickly handled the communication of the science behind the meta-analysis, thereby diverting media attention and in so doing arguably not compromising the safety of patients.

Going forward, managing risks in post-trust societies will remain difficult. There is no magical solution, no matter what some communication and risk consultancies say. Understanding the key components of trust, fairness, competence and efficiency will remain important. Risk management decision trees, as outlined on page 131, will help. What will be of special interest is managing the optimistic expectations among regulators, in particular with regard to increasing transparency in the policy-making process. Greater transparency can lead to greater public and stakeholder concern, which in turn will put strain on the communication departments of the respective regulator. In addition, greater transparency will empower the new regulator still further, increasingly allowing an organization or an individual to, in effect, set the regulatory agenda, as witnessed more and more in the pharmaceutical sector in the US (e.g. Lofstedt, 2008).

Finally, in managing risks in a post-trust society, the cultural dimension will remain important. As noted in this book, nations have different cultures, and hence different attitudes to risk; it would be wise for the European Commission and related bodies to refrain from exporting trust-building solutions from one nation to another. Notions such as critical trust may work in one society (UK) but probably not in another (Sweden).

I am grateful that Earthscan has agreed to publish the paperback version of this book. I hope it will make it more accessible to policy-makers, academics and students who have been unable to afford the purchase of the hardback version, something that has not gone unnoticed (e.g. 6, 2006). If readers have any questions regarding my reasonings and arguments made in this volume, please feel free to contact me via email on ragnar.lofstedt@kcl.ac.uk.

London
September 2008

References

6, P. (2006) 'Review of *Risk Management in Post-Trust Societies'*, *Journal of Risk Research*, vol 9, pp98–100

Barber, B. (1983) *The Logic and Limits of Trust*, Rutgers University Press, New Brunswick, NJ

House of Lords (2000) Select Committee on Science and Technology: Science and Society, The Stationary Office, London

Lofstedt, R. E. (2004) 'Risk communication and management in the twenty-first century', *International Public Management Journal*, vol 7, pp335–346

Lofstedt, R. E. (2008) 'Risk communication: The Avandia case – a pilot study', unpublished manuscript

Matsumoto, T., T. Shiomi and L. Nakayachi (2005) 'Evaluation of risk communication from the perspective of the information source: Focusing on public relation officers for nuclear power generation', *Japanese Journal of Social Psychology*, vol 20, pp201–207

Nissen, S. and K. Wolksi (2007) 'Effect of rosiglitazone on the risk of myocardial infarction and death from cardiovascular causes', *The New England Journal of Medicine*, vol 356, pp2457–2470

O'Neill, O. (2002) *A Question of Trust*, Cambridge University Press, Cambridge, UK

Peters, R. G., V. T. Covello and D. B. McCallum (1997) 'The determinants of trust and credibility in environmental risk communication: An empirical case study', *Risk Analysis*, vol 17, pp43–54

Pidgeon, N. F., W. Poortinga and J. Walls (2007) 'Scepticism, reliance and risk management institutions: Towards a conceptual model of critical trust', in M. Siegrist, T. C. Earle and H. Gutscher (eds) *Trust in Cooperative Risk Management: Uncertainty and Scepticism in the Public Mind*, Earthscan, London

Poortinga, W. and N. F. Pidgeon (2003a) *Public Perceptions of Risk, Science and Governance: Main Findings of a British Survey on Five Cases*, Centre for Environmental Risk, University of East Anglia, Norwich, UK

Poortinga, W. and N. F. Pidgeon (2003b) 'Exploring the dimensionality of trust in risk regulation', *Risk Analysis*, vol 23, pp961–972

Schütz, H. and P. M. Wiedemann (2000) 'Hazardous incident information for the public: Is it useful?', *Australasian Journal of Disaster and Trauma studies*, vol 2

UK Strategy Unit (2002) *Risk: Improving Government's Capacity to Handle Risk and Uncertainty*, Strategy Unit, Cabinet Office, London

Preface to the Hardback Edition

The aim of this book is to shed further light on how one best communicates risk in a 'post-trust society'. It is fair to say that, at least since the 1980s (if not earlier), the risk communication and management climate in a number of countries has changed tremendously. The primary reason for this change has to do with the erosion of the general public's trust toward industry and regulators. There have been a number of explanations of why the public's trust toward these bodies has decreased so dramatically, including:

- the rise of 24-hour television and internet leading to the public not having to take policy makers' comments for granted
- the concentration of political power
- the amplification of risk by the media.[1]

That said, the single most important factor to the decline in public trust has to do with the sheer number and size of regulatory scandals that has plagued Europe in particular.[2] The most significant scandals include the Belgian dioxin crisis of the summer of 1999, the tainted blood scandal in France, and the UK and European BSE crisis in the 1990s. These scandals should not be underestimated. The Belgian dioxin crisis, which involved dioxin entering the Belgian food supply via contaminated animal fat used in animal feed supplied to Belgian, French and Dutch farms, for example, had significant repercussions. Because the Belgian government did not promptly go public with the knowledge of the crisis, it was accused of self-serving cover-up leading to the resignations of two cabinet ministers and the ousting of the ruling party in a national election.[3] With regard to the BSE crisis, in which UK government ministers continued to reassure the public that BSE was not transmissible to humans even after it had begun to cross species barriers, also had significant political repercussions. John Major, the Prime Minister at the time of the BSE crisis viewed it as the worst crisis since the 1956 Suez debacle, while the then European Commissioner for Agriculture, Franz Fischler, viewed BSE as the biggest crisis the European Union had ever had.[4] In terms of decline in public trust, in a 15-year period from the early 1980s to the mid-1990s, according to the World Values Survey, the public's confidence in parliament has fallen significantly in many European countries.[5] For example:

	Early 1980s	*Mid-1990s*
Finland	65 per cent	33 per cent
Germany	51 per cent	29 per cent
Spain	48 per cent	37 per cent

In the countries where I have conducted most of my research, namely the UK and Sweden, the public's trust toward policy-makers has fallen. In the UK, polls indicate that the public's trust decreased from 39 to 22 per cent in the period 1974–96,[6] while trust in Swedish policy-makers declined from 65 per cent in 1968 to 30 per cent in 1999.[7] The issue of falling trust levels is important. First, past research indicates that it is much easier to destroy trust than to build it.[8] It is therefore highly unlikely that regulators in the UK, for example, will be able to rebuild public trust levels to the same height as they were prior to the BSE scandal, although one should note that the falling trust levels have tapered off. Second, in research that Paul Slovic, myself and others have done over recent years show that public trust is one of the most important explanatory variables of the public's perceptions of risk.[9] That is to say if the public trusts regulators, then they will perceive the risks to be less than when they do not trust the regulators. In fact, there is a correlation between low public perceived risk and a high level of public trust and vice versa. In sum, as the public becomes increasingly distrustful, the public is increasingly risk averse.

In this era of declining public trust it becomes increasingly difficult to communicate risks. This is particularly the case when public distrust is combined with elements of scientific uncertainty. Yet, that said, good risk communication is still possible. In this book, using four case studies that I have independently researched, and which I have written up and had peer reviewed and published in different risk journals, I identify a series of methods of how to best communicate risk in what I call a 'post-trust society'.

This book has been more than six years in the making. The idea for it originated when, as a Reader in Social Geography at the University of Surrey, I stumbled upon the concept of social trust and risk via the work of Professors George Cvetkovich and Timothy Earle (both at

Western Washington University) in their most interesting book entitled *Social Trust: Toward a Cosmopolitan Society*.[10] Through reading this book and subsequent weeks spent in Bellingham working alongside Professors Cvetkovich and Earle, we all felt there was a need to work collaboratively on trust. This led us to establish the Bellingham International Social Trust (BIST) meetings and the publication of our book *Social Trust and the Management of Risk* in 1999.[11] Following these meetings, I took the view that it was necessary to somehow bring together the trust literature with that of risk management so as to, in effect, reconceptualize the trust concept. In this effort I am intellectually indebted to Professor Ortwin Renn (University of Stuttgart) for both his and Levine's seminal article on trust[12] as well as his risk management/types framework.[13] The definitions of trust that Renn and Levine put forth in their article, along with his risk management framework underpins the conceptual thesis of this book. To move this theoretical project forward I requested a sabbatical from the University of Surrey, which was duly granted. For this I am extremely grateful to Professor Roland Clift, the Director for the Centre for Environmental Strategy at the University of Surrey. For the Sabbatical, I spent the academic year 1999–2000 at the Harvard Center for Risk Analysis, Harvard School of Public Health. It provided an intellectually rich environment for pursuing this work. Professor John D. Graham, the then Director of the Centre, offered insights and support at every step of the way. In addition, Professor Jim Hammit and Dr George Gray encouraged the fruition of this project, being extraordinarily generous with their time and offering fresh ideas and lessons from their own research.

Upon my return to the UK, I circulated parts of draft chapters to a number of academics most notably: Professor Asa Boholm, (Goteborg University), Professor E Donald Elliot (Yale Law School), Professor Sven-Ove Hansson (Royal Technical University, Stockholm), Dr Tom Horlick-Jones (University of Cardiff), Dr Joanne Linneroth-Bayer (International Institute for Applied Systems Analysis), Edward Taylor (*Wall Street Journal Europe*; he also thought up the title for the book), Professor Kip Viscusi (Harvard Law School), and Dr Perri 6 (University of Birmingham) all of whom provided useful comments. Following an appointment to King's College and the establishment of the Centre for Risk Management, I was forced to put the book in a desk drawer where it more or less remained until January 2004 when the book was redrafted. The rewriting of this book was most ably aided by two expert editors at King's College, namely Howard Fuller and Rebecca Oxley, both of whom polished the English and sharpened some of my intellectual arguments.

The completion of this book would not have been possible without the support of a number of funding bodies most notably: the Centre for Technology Assessment in Baden-Wurttemberg, Harvard Center for Risk Analysis–Harvard School of Public Health, Pfizer Global Research and Development, the Swedish Council for the Planning and Coordination of Research, the Swedish Research Foundation, the UK Health and Safety Executive and the University of Surrey.

Throughout my years as an academic, I have been supported by three mentors, namely: Professor Baruch Fischhoff of Carnegie Mellon University, Professor John Graham, now at the Office of Management and Budget, and Professor Ortwin Renn, at the University of Stuttgart. With regard to this book, they have offered intellectual encouragement, reading numerous drafts and commenting on a wide array of intellectual ideas, as well as pushing me to complete the final draft and subsequent publication. To the three of them, thank you! Finally, the completion of this book would not have been possible without the support and love of my wife, Laura, and my two daughters, Anneli and Åsa, to whom this book is dedicated.

RAGNAR E. LÖFSTEDT

List of Abbreviations

BATNEEC	Best Available Technology Not Exceeding Excessive Cost
BPEO	Best Possible Environmental Option
BPM	Best Practicable Means
BSE	Bovine Spongiform Encephalitis
CLF	Conservation Law Foundation
DNV	Det Norske Veritas
DTI	Department of Trade and Industry
EPA	Environmental Protection Agency
FERC	Federal Energy Regulatory Commission
GAO	General Accounting Office
HSE	Health and Safety Executive
IP	International Paper
MAFF	Ministry of Agriculture Fisheries and Food
NERC	National Environmental Research Council
NGOs	non-governmental organizations
NRC	National Research Council
OMB	Office of Management and Budget
OSHA	Occupational Safety and Health Administration
OSPAR	Oslo-Paris Convention
PAN	*Gesellschaft zur Planung der Restabfallbehandlung in der Region Norschwarzwald* (group responsible for the planning of the North Black Forest's Wastes)
RCEP	Royal Commission for Environmental Pollution
SAB	Scientific Advisory Board
SAMS	Scottish Association for Marine Science
SKI	Swedish Nuclear Inspectorate

1
Introduction and Overview

Risk communication helps companies, governments and institutions minimize disputes, resolve issues and anticipate problems before they result in an irreversible breakdown in communications. Without good risk communication and good risk management, policy-makers have no road map to guide them through unforeseen problems which frequently derail the best policies and result in a breakdown in communications and a loss of trust among those they are trying hardest to persuade. Most policy-makers still use outdated methods – developed at a time before health scares such as BSE, genetically modified organisms and uranium-tipped shells eroded public confidence in industry and government – to communicate policies and achieve their objectives. Good risk communication is still possible, however. In this book, through the use of a host of case studies, I identify a series of methods that are being used in a post-trust society. That said, there is no such thing as a formula for risk communication. The same risk communication strategy may have different outcomes depending on the audience, the country, and context in which it is used. A strategy for managing risk in the USA, for example, may be wholly inappropriate in a European context.

Thanks to good risk communication, the Swedish Chemical Inspectorate has managed to retain industry and public confidence while implementing one of the strictest regulatory regimes in the world. Since 1992, the Swedish Chemical Inspectorate has implemented one such regime regarding the use of antifouling paints on pleasure boats. The Inspectorate managed this despite initial opposition from the pleasure boat owners themselves, the various national boat and yachting clubs, as well as the manufacturers of antifouling paints. In its two rulings thus far (1992 and 1998), the Inspectorate has denied approval for the use of these paints in any fresh water area as well as along the entire Baltic Sea coast for environmental reasons. The overall aim of the Inspectorate is to phase out usage of all marine craft antifouling paints, from pleasure boats to working vessels. Since there was less need for such paints on the former, they decided to concentrate on them first. Anticipating an outcry from the pleasure boat owners and the yachting clubs that represent them, the Swedish Chemical Inspectorate invested some time and money communicating the risks of antifouling paints to pleasure boat owners and their organizations. For example, in its communications the Inspectorate argued that antifouling paints are poisonous; they contain toxic chemicals and heavy metals which kill algae, blue mussels and plankton.

The main aim of the risk communication programme was therefore to help ensure a smooth passage for the antifouling legislation and to attain maximum compliance among the affected boat owners. This risk communication initiative had three major components:

(a) meetings with representatives from boat owners' associations and other government bodies to discuss changes in legislation regarding use of antifouling paints;
(b) dissemination of information regarding the proposed phasing-out of toxic paints in Sweden aimed at the boat owners themselves (for example, information stands were erected at local and nationwide boat shows such as the annual Älvsjömässa);
(c) use of the media to explain the reasoning behind the legislative changes.

The results of this exercise indicated that the initial dialogue with the boat-owning associations was not successful. The representatives questioned the science behind the Inspectorate's decision, and argued that there were no realistic, cost-effective alternatives in place, and also that phasing out the antifouling paints would not necessarily lead to a better environment. In addition, the boat-owning organizations pointed out that phasing out antifouling paints would ultimately not work as boat owners would circumvent the legislation by buying such paints from outside Sweden or via the Internet.

The second component of the risk management strategy, the communication with the leisure boat owners themselves, was highly successful, however. A majority of the boat owners interviewed favoured the Chemical Inspectorate's decision not to renew the licence for antifouling paints along the Baltic Sea either because they felt that there were viable alternatives or because there would be environmental benefits in doing so. In effect, the boat owners upheld the decision made by the Inspectorate.

Conventional wisdom dictates that stakeholder dialogue is the best way forward. However, this case illustrates how stakeholder dialogue does not always lead to the best risk communication strategy. It shows that when confronted with a conflict composed of 'low uncertainty' and complexity (biocides are known to kill algae, among other things), and when there is trust in the regulator, a top-down form of risk communication (information transfer) may be better than dialogue. Indeed, the dialogue process in this case led representatives of the boat-owning organizations to question the process of creating the legislation. In so doing, the boat-owning organizations tried to portray the Inspectorate as untrustworthy to the boat owners. Had this 'counter' risk management strategy been successful in amplifying distrust, the overall chemical risk communication programme would have backfired.[1]

This book focuses on these sorts of issues. It argues, contrary to popular and political opinion, that dialogue risk communication and stakeholder involvement in the policy-making process is not the be-all and end-all of risk management.[2] One needs, rather, to proceed on a case-by-case basis and to test for trust, since whether each party (industry or regulator) is trusted or not will determine what type of risk management strategy should be implemented. When one is not trusted one has to examine why, and then factors such as fairness, competence and efficiency come into play.

The book is divided into seven chapters. This first chapter focuses on the meanings of trust and risk management and concludes by outlining a series of risk management factors which will be examined in more detail at the end of Chapters 3–6. Chapter 2 reviews four risk management strategies (a political regulatory process including litigation; public deliberation; technocratic/scientific perspective; and risk management on strict economic grounds), and summarizes in detail the background literatures on deliberation, competence and efficiency. Chapters 3–6 are case studies taken from four different countries in which the various risk management tools summarized in Chapter 2 are applied and evaluated. In the concluding chapter, lessons for theory and practice are discussed, including an evaluation of some of the factors that influence risk management and the development of a risk management 'decision tree'.

Why this line of inquiry?

Public trust in policy-makers, industry officials and opinion-shapers is declining in western societies. Opinion polls in countries such as Sweden, the UK and the USA all highlight this.[3] This decline in public trust appears to be related to a number of factors, including social alienation; a lack of social capital; higher levels of education and greater availability of information resulting in a more sceptical public; increased scientific pluralism leading to confusing messages; cronyism in government; growth of citizen activism in an era of complex and uncertain risks and multiple messengers; regulatory scandals, such as contaminated blood in France and BSE throughout Europe; and a hyper-critical media.[4] Risk managers are increasingly aware of these factors. Corporations such as Monsanto and Shell, both having scored own goals in recent years, are nothing less than obsessed by them. The public's lack of trust in regulators, industry officials and opinion-formers is manifested in the difficulty of siting and building industrial plants, whether with or without environmental impact, such as chemical plants and wind power parks, as well as in the difficulty of disposing of obsolete plants (e.g., the oil storage buoy called Brent Spar: see Chapter 6).

Risk managers have, over time, re-developed a series of tools or techniques for dealing with these types of problems.[5] The term 're-developed' is used as most of these techniques – such as expert-based (technocratic) risk management – predate the decrease in public trust of risk managers. Hence risk managers are trying to re-develop these tools today, to make them more suited to managing risks effectively and efficiently while at the same time maintaining public trust.

Risk managers are concerned about declining public trust, which they regard as jeopardizing the efficiency of the risk management process. Indeed, the use of some of these tools, depending on the cultural setting, was nullified by the public's mistrust, and in some cases actually increased distrust. This presents distinct problems for risk regulators:

(a) it is much easier to lose or destroy trust than to gain it;[6]
(b) in an 'era of distrust', the public will turn to other sources of information and (in many cases) believe this information more than that provided by the public risk managers, even when the latter may be more accurate;
(c) when the public has access to many more sources of information, such as the Internet and 24-hour television, they are no longer dependent on policy-makers or risk managers for information. The result is a more knowledgeable but more sceptical public.[7]

A debate now exists about the best way to deal with this decline in public trust. One view is to decrease public involvement, taking the view that the public already has too much influence on risk regulation, resulting in both the wrong types of problems being prioritized and inefficient decision-making.[8] Other risk managers and their advisers argue that only by increasing public deliberation, from as early as the risk characterization stage, can public trust be increased, leading to less public opposition regarding the regulation/measure put in place.[9]

There is, however, relatively little discussion of the merits and deficiencies of these proposed strategies in increasing public trust or whether they are either culturally or geographically dependent. On this last point, for example, one could hypothesize that technocratic risk management would fit better in Sweden, where the political culture is centralized, homogeneous, and elitist and where conflict is avoided than in the USA, where the policy-making process is decentralized and debate is encouraged.[10]

In offering solutions to the present regulatory dilemma caused by public mistrust, the research is based on the following premises.

1 *Regulation is essential*: it offers advantages for efficiency as well as equity. It holds obvious benefits for both public health and the environment.[11]
2 In order to be effective and influential, *regulatory bodies need to have public trust*.[12]

3   There is evidence that *public trust in regulatory bodies is vulnerable, uneven and may be declining overall.*[13]
4   Therefore, *there is a need to re-examine the use of the various risk management tools* to help buttress public trust in regulation, but without inducing a false or unwarranted degree of confidence in public management.[14]

Trust

Trust provides us with the lubrication to ease inherent frictions between society and its regulators. This concept, as defined below, helps us explain why public and interest groups[15] have more confidence in some risk management strategies (partially dependent on their cultural and geographical background) than in others.

So what exactly is trust? In a recent literature review, Kramer and Tyler noted that there are no fewer than 16 definitions of the word.[16] Trust can be an expression of confidence between the parties in an exchange transaction and can either be process/system or outcome-based. This book argues that it is both. In some cases, for example, the public will trust regulators even if they do not agree with a regulatory decision, as long as they understand the process itself to be credible (i.e., fair, competent and efficient). In most cases, however, the public judges regulators on their past decisions (outcomes). If the public perceives the regulator to be competent, fair and efficient (the so-called three dimensions of trust) based on previous decisions, the public is highly likely to trust these regulatory bodies in the future. This approach to trust would seem the most useful. As Anthony Giddens argues: 'Trust may be defined as confidence in the reliability of a person or system, regarding a given set of outcomes or events, where the confidence expresses a faith in the probity or love of another, or in the correctness of abstract principles (technical knowledge).'[17]

For our purposes here this rather general definition will need further refinement. Trust can be seen as a 'complexity reduction thesis': that is, 'trust' means acceptance of decisions by the constituents without questioning the rationale behind them. In such a case constituents are in effect asking to accept a 'risk judgement' made by the regulators.[18] In using this definition, trust becomes something regulators should strive for. It is always easier to trust than to distrust. In order to understand how the three dimensions of trust (fairness, competence and efficiency)[19] affect modern day risk management, it is useful to look at them in more detail.

Fairness

Impartiality and fairness (also one of the main factors of deliberation) are important elements of any regulatory decision that will have an impact on public trust, and are cornerstones of a just society. There are two ways to measure fairness in regulation: either via the process itself or through the outcome of the process. Fairness is usually defined by a view of the process or outcome as being impartial. Did the regulators take everyone's interests into account, not solely those of certain powerful industrial bodies? If the regulators are not seen as impartial or fair they are unlikely to gain trust. In such cases deliberative mechanisms including public or interest groups may be needed to build trust. For local disputes, public participatory measures are appropriate, but for national – and in particular international – disputes, interest groups need to be involved. It must be said that involving interest groups in the risk management process is a high risk strategy, as they may seek to unduly influence a process to fulfil their own ends, and there is therefore no guarantee of a successful outcome. Careful management of the process is essential.

Competence

Public perception of risk managers' competence (one of the underlying variables of technocracy) is viewed by researchers as the most important component of trust.[20] The easiest way to measure regulators' competence in a specific process is to evaluate it. Did the regulators handle the process as proficiently as possible? Did the risk managers have the necessary scientific and practical background to deal with the range of issues associated with the process? If the regulators are not seen as competent, thereby compromising trust, additional expertise may need to be brought into the process (e.g., scientific advisory boards).

Efficiency

The third component of trust is efficiency and this can be viewed as how taxpayers' money is used in the regulatory process (saving lives or safeguarding the environment).[21] The efficiency argument is particularly important during periods of economic stress, when levels of government expenditure have significant effects on the public's welfare and state of well-being. The concept of relating efficiency to trust is underdeveloped because in many cases what the economists and/or technocrats see as inefficient, such as spending public funds on cleaning up contaminated land sites (e.g., the USA's Superfund project), is seen by the public as very important for reasons other than efficiency.

*   *   *

These three trust factors are closely tied to the three risk management tools that are highlighted in this book, namely deliberation, technocracy/expert and efficiency/rational. They will be referred to frequently over the next few chapters.

The role of trust in risk management

Only in the last 20 years have researchers focused their attention on the link between public trust and the management of risk in North America and Europe. The reasons for this research field being explored so late are outlined below:

1   Risk perception itself, as applied to technological hazards, was not popularized until the late 1960s. The research area was dominated by the so-called 'psychometric paradigm' in which psychologists focused on the reasons/factors why people were concerned about some risks more than others.[22] Hence, because the risk research field was only discovered later, it should not be surprising that the risk perception trust link was not popularized until recently.

2   In the USA, with a dynamic culturally-oriented competitive climate and a political tradition of 'checks and balances', there was little need to examine the role of trust in the risk management process as the risk management regulatory system was inherently based on distrust.[23] In effect there were elements of distrust in most of the bodies involved with regulation. The regulators would sue industry for not complying; industry would sue the regulators as they found the regulators too tough, and environmental groups would sue the regulators for being too lenient on industry and industry for not complying.

3 In Europe, the opposite was the case. Regulators had no reason to believe the public would *not* trust them; the issue hardly arose since regulation was based on some form of consensus in which the regulators and industry were seen to have the public's best interests at heart.[24]

Researchers, as well as risk management practitioners, took an interest in the field in the 1980s and 1990s when they started to find correlations between high public perceived risk and high distrust, and vice versa.[25] Regulators realized it was now in their best interest to build up trust with the general public. Doing so would reduce conflict, leading to the regulatory process becoming more efficient.

Following the publication of these findings, risk researchers regarded trust as one of the most important dimensions for understanding publicly perceived risk. These research findings further increased in importance for risk managers when European pollsters in the mid to late 1990s showed a decline in public trust towards government institutions. Based on the public's distrust of policy-makers, opinion formers and industry, it is not surprising that the public increasingly felt unsafe, although society on the whole has never been so risk free, due to a series of factors ranging from pasteurization of milk to clean drinking water and the invention of penicillin.[26]

This prompted researchers to examine how public trust in industry and government can be built up. In many cases, the findings made stark reading. In a survey of the US nuclear industry, for example, researchers concluded that because it systematically misled the American public for so long, the industry deserved to be mistrusted. With such a legacy of deception it was unlikely public trust towards this sector would increase in the foreseeable future.[27] Other researchers pointed out the reverse, however: no matter how great the level of distrust, through the involvement of the public and interest groups (e.g., via deliberation) in the policy-making process it is possible for public trust in policy-makers to increase.[28]

In summary, for risk managers the research showed that there was a link between high perceived risk and high distrust (and vice versa); and also that regulators (particularly in Europe) were trusted less and less. Based on these and other related findings, regulators started taking the whole area of risk communication, public perception of risk, and the role of trust in the management process much more seriously.

The conceptual ideas of this book

Conceptually this book tries to develop some of the theoretical risk ideas present in both the trust and risk management literatures. For the sake of argument these ideas or risk factors are grouped under the following headings:

- context of the decision-making process
- behaviour of the risk managers
- perception of the actors involved

The ideas and factors can be summarized as described below.

Context of the decision-making process

In a high public trust, high/low uncertainty risk situation, deliberative risk management strategies are not required

Deliberation, if used incorrectly, may in effect increase public/ stakeholder distrust in the policy making process. This argument goes against many of the current views held by research bodies and institutions such as the European Commission. These bodies take a carte-blanche view arguing that deliberation is in one way or another always good. I argue, however, that deliberation works best in situations when the public does not trust policy-makers, opinion-formers or industrialists because they are seen as unfair or partial. Deliberative techniques are not useful when there is already high public trust in the regulatory process, even when decision stakes and uncertainty levels are high. In fact, deliberation in such a situation may in effect lessen public trust of the regulator. In such cases, regulators should refer to experience: that is to say, they should implement the risk management strategies they have used previously, as these strategies in the past have led to public trust.

In a low public trust situation, a risk management strategy (strategies) will need to be implemented, but the strategy selected depends on the reasons for the distrust in the first place

As stated earlier there are three dimensions of trust: if the public sees the regulator as incompetent, some form or shape of expertise (technocracy) is required. If regulators are seen as unfair (partial), then some form of deliberation is required. If the regulator is seen as inefficient by the public, then rational risk (economic) mechanisms are necessary. In effect, the regulator/industry in question should 'test for trust' and, if distrusted, uncover why and act appropriately.

Deliberative techniques can help create public trust regarding a contentious risk management issue, if public distrust has something to do with partiality, but are expensive and time-consuming

In cases where the public perceive the regulators as partial, one of the more effective (yet most expensive) ways to increase public trust in the policy-making process is to include them in it. Deliberation may work under these specific circumstances, particularly at the local level where the public is involved (and when interest groups are kept out), assuming that the outcome of the process is adopted.

Behaviour of the risk managers

In high distrust situations, charismatic individuals are extremely helpful in negotiating successful deliberative outcomes

In many cases charismatic individuals can make or break outcomes. This may seem to be an obvious conclusion for risk managers involved in the regulatory process. That said, often regulators have not attempted to involve such charismatic individuals, either because they are not seen as necessary, or because, for one reason or another, the charismatic individuals do not want to become involved in the policy-making process in the first place.

In any regulatory/risk management process the political actors, be they local or national, have to support the final outcome

If this is not the case, then the final outcome of the process may be worse than if nothing was accomplished. If an agreement is made via the involvement of interest groups, experts or the public, and if the process is acrimonious and contentious, then the political actors should be publicly supportive of the outcome of the process to help heal possible divisions among various actors, thereby unifying the community. If the policy-makers, after an acrimonious and contentious struggle, do not support the outcomes of the risk management process, the wounds can remain and the community may stay divided, thereby leading to greater public distrust in the process as a whole. Arguably, the public distrust in the divided community may then be stronger at this final stage than before the risk management process even started, as the process led to seemingly false expectations.

It is not enough to assume the regulator has public trust; the regulator also has to test and see whether there is public trust

Some regulatory bodies assume they have trust, while others (including several US regulatory agencies) assume they are not trusted by the general public and/or industrial interests.[29] The issues are never as clear cut as this. In situations where regulators assume they are not trusted, and when in fact the opposite is the case, risk managers may involve interest groups in the policy-making process, thereby complicating it. To cope with such situations, regulatory bodies should test, preferably via face-to-face, in-depth interview surveys, on a frequent basis to see whether they have public trust; and, if they do not have it, they should try to discover why not.

Proactive regulation is more likely to gain public trust

Proactive regulators, who act before a crisis is at hand, can increase public trust. Those regulators who act retrospectively and end up 'fire-fighting' with a wide array of interest groups will only increase public distrust. In acting thus, a policy vacuum occurs.[30] In many cases this vacuum is quickly filled by interest groups who will pursue their own agendas, possibly undermining regulators and thereby creating public distrust.[31]

Perception of the actors involved

Interest groups will in many cases try to create public distrust of regulators which in turn can lead to failures of the risk management process

Related to the above, interest groups by their very nature want to promote their own interests (whether halting the construction of a waste incinerator or protecting the environment in general), and the most effective way to do so is to promote distrust in those institutions that are in charge of the issue in question (usually regulators and industry). In many cases, the less interest group involvement, the more successful the outcome of the proposed strategy will be.[32]

Interest groups are needed, however, when the regulator is not seen as impartial and when one is dealing with national or international regulatory issues

In some cases the regulator is not impartial. In such cases the public or interest groups are needed to participate in the regulatory process to ensure that it is fair. With regard to local regulatory issues it is usually enough to have some form of public participation (be it citizen panels, juries or advisory boards). However, when dealing with national or international regulatory issues in which the regulator is seen as unfair, interest groups will be needed even if they may actually promote more distrust in the risk management process, as the involvement of the public on national or international issues is simply impractical. It should be made clear that involving interest groups in these cases is a high risk strategy. Public trust may be created, but may also be destroyed. The alternatives, however, are no better. If interest groups are not asked to be involved, they will inevitably choose to be so anyway with results leading to predictable distrust.

* * *

These conceptual ideas will be discussed here with regard to four case studies. Prior to this a review follows of the four major intellectual perspectives in the field of risk management which underpin this book's conceptual foundations.

2
A Review of the Four Risk Management Strategies

Introduction

Risk management encompasses a series of strategies or models. Max Weber, for example, defines four risk management 'ideal types':

(a) political regulatory process, including litigation;
(b) public deliberation;
(c) the technocratic /scientific perspective;
(d) risk management on strict economic grounds.[1]

These ideal types can be represented graphically (see Figure 2.1). This graphic illustration originates from Parson's description of society,[2] which was then developed and refined by Ortwin Renn in a number of articles in the 1990s (the one published in German in 1996 is the most significant).[3]

The four regulatory 'ideal types'

These 'ideal types' are customized to fit individual nations' needs but have different histories and came into the risk management fold at different times. Of the four ideal types the latter three diverge from the first, the political regulatory process. This is the standard form of making regulations and is composed of a number of stages including: agenda setting, decision-making, implementation and evaluation of the regulation. The political regulatory process is centuries old, having been implemented in different ways in western nations.

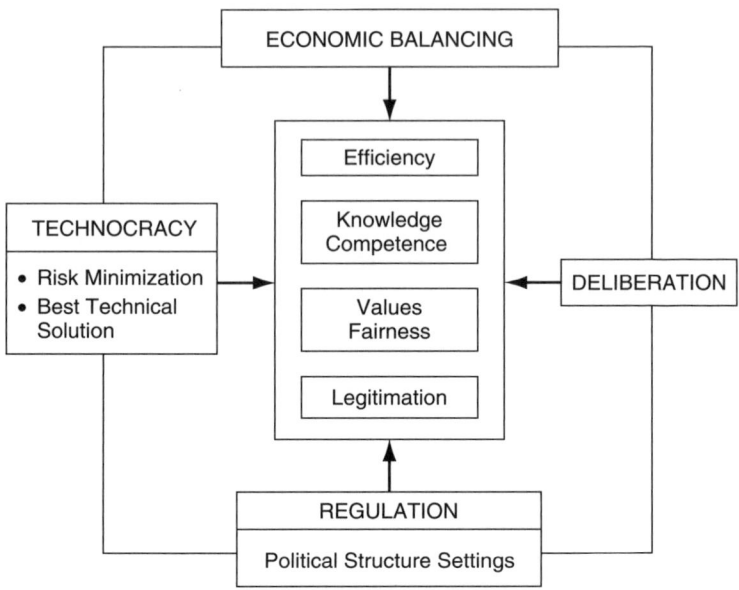

Figure 2.1   The four ideal types of risk management (© Ortwin Renn, reprinted with permission)

What happened over time, however, was that the standard political regulatory process became increasingly questioned. The US public, concerned about regulatory capture, for example, began showing less confidence in technical or scientific elites.[4] Likewise, following a number of scandals throughout Europe, Europeans began trusting regulators less and less and environmental NGOs more.[5] On the other hand, some academics argue that present day regulators are not aware of risk-versus-risk issues in setting standards[6] and that the regulatory process should therefore be more scientific. Others point out that the present regulatory environment is simply too costly, whether in terms of the electromagnetic fields in Sweden[7] or cleaning up Superfund sites in the USA.[8] These and other academics, policy-makers and a variety of stakeholders, converging from many viewpoints, began proposing different forms of regulatory strategies or ideal types borrowed from different societies and cultures. As discussed above, these ideal types can be grouped in three separate categories and they each have very different histories which have shaped their use in present day regulation.

Public and stakeholder deliberation

History

As long ago as ancient Grecian times, citizens participated in policy-making. Similarly, direct democracy rather like the Greek model was also practised in a few small cantons in central Switzerland in the thirteenth century.[9] In addition, there were signs of citizen involvement in the policy-making process in the early Renaissance in the free 'city-states' of Italy (such as Venice).[10] It was not until the Enlightenment period, however, that the fundamental elements of democracy, division of power and equal opportunity for political action (such as voting and running for political office) were articulated.[11]

After the revolutions of the late eighteenth century in America and France, public involvement in the policy-making process became firmly established in both western Europe and North America, in theory if not always in practice. Lowi, for example, argues that participation has been a recurrent theme in American history, with demands increasing over time as the role of government expanded.[12] Referenda, for example, a formal interest representation at the national level, developed in the early part of the twentieth century.[13] Following the Second World War, participation in the policy-making process continued to grow after the Administrative Procedure Act was passed in 1946. This called for due process and gave the public greater rights to comment, and gave opportunities for hearings. The 1966 Freedom of Information Act granted similar privileges. These Acts are termed the 'old' school in which participation was seen as a privilege, with only the organizations which had better resources being able to partake.[14]

The 'new' school of public participation arose with the federal administrative process of the 1960s and 1970s.[15] During these years wide public participation was seen as a necessary element in many federal statutes, and was viewed as an important contribution to democracy and to the quality of the decision-making process itself.

In Europe, deliberation grew in prevalence in the 1970s when it was used as an aid to urban planning in many communities.[16] The purpose here was to use deliberation techniques to better understand (and incorporate into the decision-making process) public values and preferences.[17] This work led to the development of various public participation techniques ranging from consensus conferences, to advisory panels and citizen juries.[18]

Present day deliberation

In this book deliberation is used to refer to an exchange of ideas between the public and interest groups with policy-makers and industry representatives. More specifically, it refers to the involvement of the public and various interest groups in a multilevel framework which characterizes the risk to be managed.[19]

Deliberation has four main purposes:[20] normative democracy, equity and fairness, more effective risk communication, and relativism of knowledge.

Normative democracy

Some advocates see deliberation as a good 'democratic value', regardless of whether it fosters competent or trustworthy decisions. It breeds better citizens, gives an affirmation of belonging and is necessary for overall system stability.[21] In effect, participation can enhance the responsiveness and legitimacy of public institutions. Otherwise, Harvey Brooks argues, the 'modern nation risks being no longer recognizable as a democracy, either representative or plebiscitary, if more and more areas are excluded from public participation because of technical complexity'.[22] Or, as one of America's 'Founding Fathers', James Madison, emphasized in his *Federalist Paper* No.10, public participation is essential for sound government.[23] These advocates take the view that it is necessary to have public participation to keep democracy alive. Referenda (most popular in California and Switzerland) or 'New England Town Hall meetings' are examples of using the public participation approach as a way of engaging in *democracy*.

Equity and fairness

Another purpose of instigating deliberation is for equity and fairness. Public participation levels the playing field by providing citizens with an opportunity to influence their representatives. This is perhaps more of an ideological perspective, however, and its proponents argue that in a naturally uneven capitalist society risks are affect poor people proportionally more.[24] Through public participation, the public can help influence the distribution of wealth. Deliberation is therefore needed to prevent capitalists from exclusively directing public goals. Capitalists, they argue, cannot tell us what is an intolerable risk. With deliberation, the affected public can help decide which burdens are tolerable. In this process people's values and the concept of fairness get due consideration.[25]

More effective risk communication

In both Europe and the USA, using deliberation did not become popular in the eyes of risk managers until the advent of dialogue risk communication in the late 1980s, following a significant amount of research by UK and US based academics.[26]

In the area of risk communication, practitioners as well as researchers began discussing the need for dialogue, or reciprocal risk communication, in the late 1980s.[27] Contemporary studies indicated that the most common form of risk communication, 'top-down', was not successful in alleviating public fears. The public simply did not seem to be influenced by top-down risk communication programmes.[28] The main reasons identified for this failure were poor communication among the experts themselves, and their contemptuous response to public opposition.[29] This contributed to increased public mistrust.[30]

The development of dialogue risk communication techniques was welcomed among both industry and regulators, especially in the USA. Industry and local and federal government regulators, frustrated by the difficulties of siting plants and dumping and burning wastes, were keen to learn how to increase public trust via more active engagement, as well as gain information on the affected citizens' preferences by involving them directly in the policy-making process.[31] *It is because of its perceived ability to increase public trust that dialogue risk communication is so very much in vogue.*[32] Public/interest group participation is identified as important in rebuilding the legitimacy of the decision-making process[33] and has prompted an expansion of university-based centres on risk communication and consultancies. In Europe particularly, there is a rush by industry to establish deliberative risk management following the 1995 Brent Spar crisis (see Chapter 6).[34]

Relativism of knowledge

Another impetus to the establishment of deliberative procedures came from the relativism of the knowledge area, where researchers saw deliberation as a strategy to increase scientific knowledge. Academics, including sociologists Alan Irwin and Brian Wynne, questioned the expertise of scientists and other so-called experts on complex, highly uncertain hazards such as BSE and nuclear power. They pointed out that in many cases the affected publics had more knowledge about the causes, consequences and effects of a certain hazard than the experts.[35] Others argued that the use of public participation techniques supplied decision-makers with more comprehensive information on social values[36] and was critical in bringing about a re-examination of the problem definitions underlying American politics.[37] In effect, the work emphasized the importance of taking anecdotal information into account.

A widely reported study was Wynne's on Cumbrian sheep farmers who had seen their sheep contaminated by radiation after the Chernobyl accident. His research showed that the procedures for the affected sheep provided by scientists at the Ministry of Agriculture Fisheries and Food (MAFF) and other government bodies did not consider the sheep farmers' local knowledge. Instead these scientists expressed a high degree of scientific certainty about the declining levels of radiation in the sheep in the medium to long term. As a result farmers, who felt that MAFF's advice was wrong, found themselves unable to argue against scientists on 'merely' a common sense approach. Time went on and radiation levels in the sheep did not decrease as the scientists confidently predicted; their own credibility, however, did. It also became evident that the farmers' knowledge was more useful in explaining why levels of radiation did not decline. Wynne therefore concluded that the public and other stakeholders should be more involved in risk management decisions.[38]

Different forms of deliberative techniques[39]

Deliberation comes in different forms. There are citizen advisory boards where selected citizens are asked to participate and advise on a risk management process; citizen panels and citizen juries where citizens are randomly selected to participate in a risk management decision process which may adversely affect them (e.g., building a waste-to-energy incinerator in the North Black Forest: see Chapter 3); or industry may choose other routes such as initiating dialogue processes with influential interest groups as a part of the risk management process.

Disadvantages of the deliberative approach

Rossi[40] and others (particularly Coglianese) suggest that the deliberative process (at least in the USA) is a relative failure. There are several reasons for this. Research has shown that more deliberation in the policy-making process leads to public mistrust of policy-makers and the government as a whole; they only become more aware of how staid and ineffective bureaucrats can be. Some scientists and experts criticize deliberation, as they see no reason why the public, whom they view as under- or misinformed, lost, bewildered, overtly self-interested or simply apathetic, should participate in policy-making[41] at all. In addition, the popularity of the deliberative process in the 1990s has resulted in more work for government officials and a less efficient practice. Bureaucrats concerned about transparency have now begun to retreat from public meetings, favouring memos and informal discussions.[42] The participatory process is also criticized for requiring the public to assimilate a lot of information about the particular issue being deliberated, which can lead to divergent rather than convergent views.[43]

Technocracy and rational risk: symptoms of centralization

History of the technocratic approach

It is difficult to determine the exact history of the technocratic approach. Its proponents take the view that risk management should be left to the elites, led by science and strict peer review. Thus its origins, it can be argued, are that of advising regulators. It can, for example, be seen in the civil servants in France and the UK.[44]

Viewed this way, however, technocratic risk management strategy is no different from the political regulatory process described previously. Both have elites and both make regulations. It has been argued, however, that technocracy is based on expert involvement in the policy-making process, referred to as 'expertocracy'. From this perspective the field has its origins in the establishment of expert advisory agencies and councils. Although these agencies and councils existed in some nations for many years[45] they did not become very popular until the 1970s, with the advent of social regulation and the formation of a wide range of federal regulatory agencies such as the Environmental Protection Agency (EPA). At that time these agencies were forced to make explicit trade-offs between regulations' economic costs and risks to environment and public health. This was fuelled by a number of controversial decisions in the chemical and public health area ranging from Love Canal to 2,4,5-T herbicide regulation.[46] In the UK, USA and elsewhere, the use of expertise in and outside government has grown tremendously, as scientific pluralism has increased due to lack of public trust towards regulators and increased scientific uncertainty regarding minute risks (e.g., genetically modified organisms and radiation from mobile telephone handsets).[47]

Experts advising the various federal agencies can be inside or outside government. For example, the EPA has its own Scientific Advisory Board (SAB), staffed largely by academics. Arguably the SAB should be seen as working within government, and funded by government. There are also experts outside government who provide advice to the EPA, which they may or may not take. These experts may be independent academics or staff members of environmental groups that may criticize an EPA decision (and perhaps even sue the agency) or provide alternative advice for a fee or free of charge. Those experts that work within government, such as senior scientific civil servants or members of the EPA's SAB, are referred to here as the 'technocracy'.

Current use of technocratic risk management[48]

Proponents of the technocratic perspective feel risk management should be left to the elites/experts advising government ministers and policy-makers with minimal or no public involvement. Only through strong science-led, expert advice and strict peer review will risk management ultimately work. Technocrats/experts want risk managers (civil servants) to create outcomes that citizens, after careful deliberation and training in relevant sciences, would want the government to produce.[49] They see themselves as delegated agents of lay citizens who lack the time, expertise, resources and cognitive capacity to make complex risk-management decisions.

Notions of fairness as well as efficiency are important for technocrats. Technocrats are sceptical of stakeholder-based decision-making as well as decisions based on opinion polls and/or raw popular opinion. Involving the public and interest groups in a deliberative fashion can lead to inefficiencies both in time and funds, wrong prioritization of the hazards to be managed, and unforeseen difficulties, all of which breed distrust. By leaving risk management to experts, who know the issue better than anyone else, society benefits.[50] Nevertheless, technocrats argue that some form of public participation is needed to ensure accountability, and to force technocrats/experts to formulate decisions that are understood by the public.[51] The technocrats are experts. They know the area that they are set to regulate better than any other. They serve as advisers to civil servants and ministers via expert advisory councils and agencies. They are not part of the politically appointed establishment, but are rather a politically insulated bureaucracy or expert unit that is assigned to deal with risks.

The technocratic risk management approach is well mapped out by John Graham, who argues that regulation of environmental and health problems should be based on the following criteria:[52]

- scientific expertise indicating that exposure to identified pollutants can represent significant harm to the environment or human health
- environmental problems identified should be prioritized by some type of 'comparative risk process' so as to ensure efficient use of resources
- to avoid risk–risk trade-offs, the proposed regulation should reduce the risks of the pollutants targeted to a greater extent than they increase other risks to the environment
- economic costs of the proposed actions must be reasonably related to the degree of risk reduction

In summary, regulatory reforms should be based on risk criteria drawn from economic and scientific spheres. By doing so regulators are not drawn into populist regulatory arrangements which may satisfy the concerned public, but could have negative effects on the environment as a whole. Examples of such legislation are the US EPA's strict regulatory policies on toxic chemicals in the soil and asbestos in buildings, which are arguably low actual risks, while it ignores substantial environmental problems such as indoor air pollution.[53] The technocratic approach, as advocated by Breyer, is to avoid inconsistencies caused by public and interest group opinion in the regulatory structures.

Policy-makers and regulators tend to favour the technocratic risk management perspective. It is more efficient (both in terms of time and money) than the deliberative approach and less controversial than the economic risk management alternative.[54] The technocratic approach is arguably the exact opposite of the deliberative one. Ruckelshaus, a former US EPA Administrator, argued in the early 1980s that having scientists and experts characterizing the risks and carrying out the risk assessments would restore the credibility of the US EPA.[55] In other words, cutting out interest groups and the public from the risk assessment part of the risk management process would not only lead to more efficient and competent decisions, but actually to greater public trust in the institution.

One important aspect of the technocratic approach is the risk–risk trade-off, to which policy-makers pay too little attention. Graham and Wiener postulate 'that efforts to combat a target risk can unintentionally foster increases in countervailing risks'.[56] 'Countervailing risks' can range from unintended consequences of public policy to medical side effects. To reduce the chances of risk–risk trade-offs, decision-makers need to consider all aspects of any regulatory policy. Proponents of risk–risk trade-offs note that public-driven regulatory agendas in many cases ignore the risk–risk trade-off. By adopting certain regulatory policies, risks in other areas may actually increase. A good example is the issue of water chlorination. Following risk studies in the USA which classified chlorination as carcinogenic, Peru stopped chlorinating the water in Lima in 1991, resulting in an outbreak of cholera which killed 7,000 people and affected nearly 800,000 others.[57]

Technocratic risk management is prominent in each of the countries discussed in the case studies of this book. The UK, for example, has long had a political regulatory process with a strong technocratic component and limited interest group involvement.[58] In Sweden virtually all risk management decisions are taken centrally by representatives from government, industry, trade unions and some particular groups with a reliance on expert output, but with little involvement from grass-root organizations or the public.[59] Much of the discussion about increasing the use of the technocratic approach occurs in the USA: for example, in 1983 Ruckelshaus proposed risk management based on strict scientific criteria as a necessary tool for identifying environmental threats.[60] In the early 1990s a series of bills were tabled in Congress advocating a risk-based approach for environmental decision-making. Although these bills have so far been unsuccessful, the indications are that risk-based criteria are increasingly utilized in government today.[61]

Criticisms of the approach

There is still considerable debate, however, about whether wider use of the technocratic approach will work in the USA. Researchers there state that the system is so fundamentally different, fraught with public distrust, and hampered by adversarial decision-making that it is highly unlikely that the approach could function in the same way as it has in France or the UK.[62] Others are more critical of the approach itself. Among the main problems are the following:

1 The decision-makers (i.e., Congressmen) see Breyer's perspective of the technocratic approach as arrogant, as well as inherently undemocratic.[63]
2 For the process to work, Breyer believes that trust is pivotal. This is not easily achieved in a litigious society such as the USA, where there is genuine concern about regulatory capture.[64]
3 There are limits to science: some trans-scientific issues cannot be resolved by science because of ethical, technological or information constraints.[65]
4 Although risk-based policy is entirely rational, some observers see it as practically unworkable since it involves a systematic under-estimation of uncertainty, as demonstrated by the implementation of the Toxic Substances Control Act.[66]
5 Critics see the proponents of the technocratic approach as neglecting aspects of risks related to opportunities for preventing rather than remedying or abating risks. This is problematic as the public places a high value on these risks. In other words, the process is driven by regulatory failure.
6 Scientific findings can be manipulated to fit the interests of the agency that sponsored them (e.g., the EPA's handling of the 1982 formaldehyde decision).[67]
7 Agencies have exaggerated the claims of 'pure' science.

8   Research has shown a strong correlation between agency inaction and science. This has led to the public remaining unprotected against recognized hazardous substances.[68]
9   Science-based regulatory statutes are a failure to date: fewer than 15 per cent of the necessary standards are promulgated under science-based statutory mandates. This has led to charges of scientific incompetence.[69]
10   Experts are also fallible, and in some cases there have been detrimental and fatal consequences (e.g., contaminated blood in France or BSE in the UK).[70]
11   A great deal of agency decision-making cannot be made on scientific judgement criteria alone, but must be infused with value judgements which may or may not be morally sound.[71]

Advantages of the technocratic approach

There are several distinct advantages of using the technocratic approach:[72]

(a)   it gives credibility to industrial regulation by giving it a stronger basis in science and economics; in so doing, science plays a vital role in legitimizing protective regulation;[73]
(b)   it reduces regulatory dependency on moral and/or legalistic claims;
(c)   it can block poor regulatory policies, where the costs of the regulation are significantly higher than possible achieved benefits to the environment and public health, thereby reducing costs of compliance;
(d)   it can help solve the regulatory conundrums of the present day which are by their very nature more difficult to detect.

History of risk management on strict economic grounds (development of rational risk policy)

There is a long history of taking costs into account in developing health, safety and environment regulations. In the UK, for example, a concept entitled 'Best Practicable Means' (BPM) was first used with the 1874 Alkali and Clean Air Act. Yet risk management from a rational risk perspective, adopting strict cost-benefit criteria, was not popularized until the 1970s. At that time US policy-makers saw a need for an economic oversight mechanism to ensure that the emerging environmental, safety and health regulations did not place excessive costs on the regulatees. In the USA, for example, the Nixon Administration put forward an informal quality of life review process (inflationary impact assessment), to discern the actual costs of regulations. During the subsequent Ford Administration this review process became formalized as the Council on Wage and Price Stability. This review process was more advisory than binding; regulations which ran over cost were not necessarily ended.

Next, under the Carter Administration, more oversight was placed on the regulatory bodies through the requirement of a regulatory impact analysis, implying the agencies 'had to demonstrate that the least burdensome of the acceptable alternatives have been chosen',[74] as well as the establishment of the Regulatory Analysis Review Group. Under Reagan, the Office of Management and Budget (OMB) took over the regulatory oversight function from the Council of Wage and Price Stability. In effect this increased the importance of the regulatory cost issue as OMB is in charge of setting the budgets of all regulatory agencies. Coupled with this transfer a strict benefit-cost analysis was added by Reagan's Executive Order 12291 to the process. This has set the criteria for all regulations up to the present day. It reads as follows:

Sec.2. General Requirements.

In promulgating new regulations, reviewing existing regulations, and developing legislative proposals concerning regulation, all agencies, to the extent permitted by law, shall adhere to the following requirements:

a) Administrative decisions shall be based on adequate information concerning the need for and consequences of proposed government action;

b) Regulatory action shall not be undertaken unless the potential benefits to society for the regulation outweigh the potential costs to society;

c) Regulatory objectives shall be chosen to maximize the benefits to society;

d) Among alternative approaches to any given regulatory objective, the alternative involving the least net costs to society shall be chosen; and

e) Agencies shall set regulatory priorities with the aim of maximizing the aggregate net benefits to society, taking into account the condition of the particular industries affected by regulations, the condition of the national economy, and other regulatory actions contemplated for the future.[75]

This shift in regulatory policy has resulted in increased attention to the regulatory costs placed on industry.[76]

There are similar developments in other countries. In the UK, for example, BATNEEC (Best Available Technology Not Exceeding Excessive Cost) replaced BPM in 1990, and is now widely used in the justification of environmental regulations within the Environment Agency; while in the health and safety field the Health and Safety Executive (HSE) still takes into account costs and benefits through the application of the 'as low as reasonably practicable' principle.[77] Strict economic criteria are not embraced by all regulatory bodies, however. For example, the European Union's Directorate General-Environment does not regard simple cost–benefit analysis as suitable for environmental regulation, as environmental and social values also need to be accounted for. As a way of dealing with the issue, the Commission in the summer of 2002 put forward a Better Regulation package in which the Commission advocated a greater use of Impact Assessments exploring costs and benefits as well as social and environmental values.[78]

Rational risk policy holds that there is only a limited amount of funding available for risk management and this should be used in the best possible way. It differs from the technocratic approach on two accounts. First, the economists want risk managers to create outcomes that would be created by perfectly functioning markets (if such markets existed). To them only efficiency counts and there is no room for public or stakeholder involvement. Second, the economists argue for risks to be individualized. The individual should decide whether it is worthwhile taking a risk or not. By putting warning labels on cigarettes, for example, individuals should arguably know that by smoking, they increases their chance of reducing their lifespan because of lung cancer. In a free market with warning labels, individuals can decide which risks to take and which not to. In such a society, insurance, is also a major factor. If individuals feel exposed they can take out insurance, thereby hedging their exposure.[79]

Leading proponents of this perspective include Richard Hahn, Cass Sunstein, Kip Viscusi and Richard Zeckhauser.[80] They argue that current regulations are killing people and costing unnecessarily large amounts of money. As Dana argues:

> The central thesis of the critique is that government could achieve the designated ends of environmental regulation at a much lower social cost by replacing rigid 'command and control regulation' with a more market-oriented system of tradeable pollution rights, pollution taxes, and monetary incentives for pollution prevention. Proponents claimed that market-oriented reforms would reduce industry's compliance costs and government's enforcement costs. Moreover, a market-oriented system, unlike a command and control system, would give industry an ongoing incentive to develop better pollution prevention technology.[81]

Breyer cites an illustrative example in his book *Breaking the Vicious Circle*:

> [there was] a case in my own court, United States v. Ottai and Goss, arising out of a ten-year effort to force a cleanup of a toxic waste dump in New Hampshire. The site was mostly cleaned up. All but one of the private parties had settled. The remaining private party litigated the cost of cleaning up the last little bit, a cost of about $9.3 million to remove a small amount of highly diluted PCBs and 'volatile organic compounds' (benzene and gasoline components) by incinerating the dirt. How much extra safety did this $9.3 million buy? The forty-thousand-page record of this ten year effort (and all the parties seemed to agree) that, without the extra expenditure, the waste dump was clean enough for children playing on the site to eat small amounts of dirt daily for 70 days a year without significant harm. Burning the soil would have made it clean enough for the children to eat small amounts daily for 245 days per year without significant harm. But there were no dirt-eating children playing in the area, for it was a swamp. Nor were dirt-eating children likely to appear there, for future building

seemed unlikely. The parties also agreed that at least half of the volatile organic chemicals would likely evaporate by the year 2000.[82]

The main thrust of Viscusi and Zeckhauser's argument is that both the public and policy-makers are prone to certain biases. These biases, strongly grounded in cognitive psychology (particularly the work of Kahneman and Tversky),[83] are discussed in significant detail by risk perception psychologists (in particular the work of Baruch Fischhoff, Sarah Lichstenstein and Paul Slovic in the USA, and Ortwin Renn and Lennart Sjöberg in Europe), the most common being:[84]

- an underestimation of large risks and overestimation of small ones
- greater value attached to eliminating a hazard rather than reducing the risk
- greater concern about visible, dramatic and well-publicized risks
- more concern about low probability, high consequence risk (e.g., a nuclear plant accident or a plane crash) than high probability, low consequence risk (such as a car crash)
- more concern about artificial than natural risks

These biases can lead to irrational regulation, affecting daily decisions of policy-makers. They are illustrated in the EPA example mentioned above, but they also come through many other Federal policies. Other examples are seen in the US Food and Drug Administration legislation where analysis has shown that regulation for new synthetic chemicals is more frequent than for natural ones.[85]

Biases can affect risk management policies in many ways. Irrational fears among the public caused by these biases, for example, can affect local and national policy-makers if they perceive that a particular regulation may be popular among the voting public. This was seen in Clinton and Gore's campaign promises to continue funding the clean-up of Superfund sites in 1996, even though research had shown this would not be efficient in terms of lives saved.[86]

Another issue that rational risk managers see as problematic is the intentional conservative bias in cancer risk assessments which extrapolate animal data (e.g., mice) to humans. This, according to some researchers,[87] has led to multiple errors in overestimating risk.

The basis of rational risk policy is the 90–10 principle; government regulators may incur 90 per cent of the cost to address the last 10 per cent of the risk.[88] Hence, reducing the risk of a particular problem to absolutely zero is extremely inefficient. Viscusi applies the 90–10 hypothesis to the Superfund case example. His calculations show that the first 5 per cent of expenditure eliminates 99.46 per cent of the total expected cases of cancers averted by hazardous waste clean-up efforts. The remaining 95 per cent of the expenditure leads to virtually no health risk reduction.[89] Moreover, these calculations show that the mean value of a life saved by the Superfund clean up is a massive $11.7 billion. Critics point out, however, that this calculation focuses on existing and not future risks. Superfund was put in place not to remedy existing risks, but rather to prevent potential risks by cleaning up sources of exposure before a risk is made real.

Under a rational risk policy, the cost of saving a life or avoiding an illness or injury should be the same across all government departments. When this is not the case, safety is reduced through diverting funding from effective life-saving activities to less effective ones. Some industries, due to how the public perceives them, have higher regulatory bands, as measured per lives saved, than others. The nuclear industry is notorious for putting forward regulatory measures that would cost millions of dollars per life saved, and which, if implemented, would take funding away from road or railway safety where regulatory measures are more cost-effective. In the USA, for example, regulatory measures in the traffic sector will only be implemented with a maximum cost of $3 million per life saved.[90]

The proponents argue that rational risk policies, based on economic criteria, should take precedence over deliberative procedures as bias in this can, in effect, allow people to die unnecessarily.[91] One of the fundamental reasons for the success of the rational risk approach to date has been the so-called 'no losses' phenomenon, whereby:

(a) regulatory costs are reduced (no litigation);
(b) industry does not face extensive uncertainty related to costs and environmental benefits;
(c) there is no excessive conservatism in the cancer risk assessments.[92]

Criticisms of rational risk policy

Rational risk policy has its critics as well, both in terms of the concept and the instruments used. The most common conceptual criticisms are set out below:

1 Notions of outcome equity are not considered.
2 The costs of regulation may be overestimated.
3 The process is elitist and unfair: why should economists decide what risk individuals should take?[93]
4 Cost-benefit analysis tends to underrate those risks that cannot be quantified.[94]
5 How does one monetize values in a rational fashion?[95]
6 With scientific uncertainty it is impossible to say anything concrete or quantitative about the benefits of regulation.
7 Markets are not value free; in fact, Sunstein argues for a distinction between social and market performance.[96]

The instrumental criticisms are as follows.

1 The difficulty in satisfying the methodological requirements or data demands sound and reliable comparative risk analysis.[97] It is likely that there is a causation between general and multiple environmental factors in causing cancer, for example.
2 As Pildes and Sunstein point out, Willingness to Pay (WTP) and Willingness to Accept (WTA) are fraught with problems as people find it very difficult to quantify the costs of abating various types of risk.[98] Similarly, the rational risk management approach does not take into account social values and norms, and these 'soft' issues are not easily quantified.[99]
3 Some researchers point out that economists do not have the proper tools to understand what publics' preferences are, or how to best aggregate them.[100]

The remainder of this book

Chapters 3 and 4 discuss the local case studies, namely the siting and planned development of a waste incinerator and two aerobic digesters in the northern Black Forest area of Germany and the re-licensing of several hydropower dams in Maine. Chapters 5 and 6 focus on two transnational cases; Barsebäck and Brent Spar, respectively. Chapter 7 summarizes the differences and similarities between the four case studies, and offers some lessons and suggestions for risk managers in western societies.

3
Germany and the Waste Incinerator in the North Black Forest

Overview

This case study examines the proposed siting and building of one incinerator and two aerobic waste digesters in the North Black Forest region of Germany. The risk management tool used was that of deliberation, more specifically a citizen advisory board, and is a good example of the deliberative approach, since the principal actors eventually agreed where the waste incinerator should be sited. This was no easy task. There was a deep, ingrained distrust between the public and the proposers of the two waste solutions. The public, media and the local policy-makers, moreover, were initially hostile to the use of the citizen advisory boards to help find a solution.

Introduction: the regulatory context

Germany puts a strong focus on strict political regulatory regimes with considerable litigation, to a greater extent than the other countries surveyed in this book. German industry desires regulations to be as detailed as legally possible, allowing for a minimal number of so-called bureaucratic afterthought decisions, and therefore giving it an element of predictability. Regulatory decisions are made by elites on a central or state (*Bundesland*) level. Similar to Sweden, principal actors, be they trade unions, certain favoured environmental bodies (e.g., BUND, Bundesverband fur Umwelt and Naturschutz Deutschland (Federal Association for Environmental and Nature Protection Germany)), and most importantly industrial organizations, are asked to make regulatory policy in a consensual style. It is expected that these groups will take due account of scientific factors and economic conditions.[1] The German regulatory system, as Ortwin Renn calls it, is a larger version of the American phrase 'the regulatory negotiation process'.[2]

Background

Political scientists see German regulation as neo-corporatist,[3] in which different interests manoeuvre on an elite level to promote their own concerns. The roots of the German corporatist approach can be found in post-war West Germany, where Allied (in particular American) occupiers encouraged the promotion of a 'social market economy'. In such an economy the government is in charge of social obligations, while economic issues are resolved in the market place. For this to work properly there needs to be frequent, albeit more formal and structured dialogue between industry and government, as social and economic issues are closely related.[4] A second consequence of Allied control was the decentralization of the German state, ensuring a dispersal of power (formulated by the 1949 Constitution of the Federal Republic of Germany), which must be understood as a response to the horrific experiences under the preceding National Socialist regime.[5]

If one were to characterize the German risk management system it should based on the following criteria:

(a) a federal system where power is dispersed, with consensus among the legislative and executive authority;
(b) a clear separation between risk assessment (science) and risk management (policy);
(c) an isolation of the policy-making process from public and stakeholder critics;
(d) a firm belief that all elite parties should work in the country's best interests.

Consensus among the legislative and the executive

Like that of the USA, Germany's political power is distributed. The country has a federal structure with its power constitutionally divided between state and national governments. Legislation is passed and judged at both federal and state level. German states therefore have exclusive control in enforcing regulations agreed upon nationally. In addition, there is considerable competition between the states for power, prestige and influence. The past 20 years have seen intense rivalry between Germany's richest states, Baden-Württemberg and Bavaria.

Germany has a parliamentary system similar to that of Sweden and the UK, in which the legislature is controlled by a political majority acting through a prime minister and a cabinet. Party discipline will therefore command a parliamentary majority in support of specific legislation. It has less of a hands-on role in passing regulations in comparison to the USA, however, simply because there is little competition between the various law-making bodies. German courts play no active part in the development of regulations, entering the law-making process much later than in the USA. Cooperation is the norm between the various policy-making branches. This relative stability is enhanced by the existence of a well-developed body of law governing the conduct of public authorities.[6] In addition, unlike their American contemporaries, the courts seldom play an enforcer role. Until recently, citizens and special interest groups were unable to sue regulators or industry over proposed legislations. In effect, the executive was virtually untouchable by those opposed to the regulatory process.

In Germany, legal agreements are reached through consensus between industry and the regulators. German industries demand legislative perfection, limiting the discretion of the bureaucrats. German industry is also highly export-dependent and is thus concerned about the impact of any new regulations on the economy. This 'preference for predictability' is shared by the German government, which goes to great lengths to accommodate industry. As in the UK, German regulators feel there is little need for 'policing' to ensure compliance as both bodies are working in the best interests of Germany.

Of course, the relationship is not always perfect. An example of this was the recent conflict between the ruling Socialist/Green coalition and industry regarding the proposed phasing-out of Germany's nuclear reactors. The initial reaction to this proposal was hostile from business organizations, Liberal and Conservative politicians and large utilities. Some federal states prepared for court actions, for example, which, if successful, would have led the government to pay out billions of euro. It was not until after lengthy negotiations that a nuclear consensus was reached.[7]

Clear separation between risk assessment and management

In Germany risk assessment and standard setting are entrusted to different committees, one purely scientific, the other overtly political, ensuring a clear separation between these two areas.[8] The risk assessments are carried out by technical committees, usually at the state (*Bundesland*) level. These committees gather information on the risk at hand and determine the principles of the risk assessment to be used. These agencies are functionally separate from the political wing of the ministry, which is responsible for the risk management process, ensuring some isolation from possible critics.

An isolation of the policy-making process from public and interest group critics

In Germany regulation happens behind closed doors at an elite level involving various industrial bodies, trade unions, regulators and certain principal actors. These discussions take place at the pre-parliamentary commission on law formation stage and should not be overlooked. The 1982 Chemicals Act was, for example, formed with significant involvement from the German Chemical Industry Association which worked in close collaboration with expert scientists/academics and the ministerial administrations. Negotiations were set in motion in 1980 and the law was passed by Parliament in 1982 with minimal alterations. Meanwhile, neither Parliament itself nor environmental groups played a significant role; the public, according to Richard Munch, was 'completely unaware of all pre-parliamentary negotiations'.[9]

Over the last few years, this process has changed somewhat. There is a greater mobilization of the public through various citizen groups (*Burgerinitiativen Citizen initiatives*: an example is discussed in this case), as well as other forms of social environmental movement such as Greenpeace-Deutschland which are distrustful of the present policy-making apparatus.[10] Another powerful actor in the German environmental policy-making process is the German Green Party (Bundnis90/ Die Grunen), the present minority partner in the government coalition led by the Social Democrats (SPD).

Trust in Germany

This form of regulatory process is largely successful. The public largely believes that these elites are working in their best interests, ensuring that the German public has a higher standard of living than most of its European counterparts. Although the public's level of trust in parliamentary and government agencies started to fall in the mid-1980s and throughout the 1990s, it is still high.[11]

The regulatory system has also worked well. Although the country has faced crisis in the past, ranging from widespread forest death (Waldsterben) in the 1980s, to the Chernobyl crisis in 1986 and the BSE crisis in late 2000, the incidents have been quickly contained and dealt with professionally. For example, when widespread forest death occurred in Germany, caused by sulphur dioxide and nitrogen oxide emissions in the 1980s, the regulators, in cooperation with industry, acted quickly to contain it, installing de-sulphurization scrubbers on all Germany's coal burning plants at a huge cost to the utility industry.[12] In 1986 at the height of the Chernobyl crisis, caused by the then Minister of Interior's lack of competence in addressing the public's widespread concern about the spread of the radioactive fall-out, the Federal Government made a quick and politically wise decision in establishing the Federal Ministry for the Environment, Nature Conservation, and Nuclear Safety in June 1986. In so doing the environmental portfolio was lifted from the discredited Ministry of Interior to a completely new agency devoid of public stigma.[13]

The role of the EU and the changing regulatory environment

Together with France, Germany is historically the most influential country in the European Commission. It is the economic powerhouse of Europe and successfully wields this weight to shape European regulation.[14] Being export dependent, Germany believes regulations passed in Germany which can have an impact on the German economy should also be passed by the EU ('cross-national harmonization'), thereby allowing German industry to remain competitive. There are several examples of Germany pushing its regulations on to Europe, ranging from Germany's recycling laws (DSD), to the phasing-out of lead from petrol, to EU-wide implementation of the precautionary principle; these initiatives show mixed results. Until recently a majority of the regulations passed by the EU were already being implemented by Germany. This is also changing, however, as the EU is increasingly relying on a wide array of procedural steps to encourage public participation,

self regulation and voluntary power-sharing by a number of economic actors.[15] This new European policy is very different from the
German regulatory model which is more institutionalized in nature.
Germany is trying to get around this EU policy conflict by 'developing a strategy that puts stronger emphasis on information, financial
incentives and negotiations as well on certain types of voluntary
agreements', and in so doing reaffirming the German corporatist
style of regulation.[16] Some researchers now believe Germany will not be
able to return to the consensual model which German industry is so
keen to keep. Munch, for example, argues that in an increasingly
'globalised' world, with rapid information flows, it will be corres-
pondingly difficult to make regulatory policy at such a consensual
level.[17]

The German case, regarding the use of citizen panels to help decide
where to build a waste incinerator in the North Black Forest, is an
example of such a change in regulatory policy-making. Here, the
elites in question realized that decision-making on a centralized level
would not necessarily lead to its acceptance either by local politicians
or the public at large. Local policy-makers and the public perceived
the risks to be high and the benefits low, while the experts perceived
the reverse. This was not unique. Both German and British citizens
are vehemently hostile to siting and building waste incinerators, so
much so that in Germany the government has resorted to exporting
waste to neighbouring countries such as Sweden and Switzerland.

The North Black Forest case[18]

Introduction: the public participation-citizen jury case study

Between January and June 1996 a public participation panel project,
a form of deliberation, was undertaken in the North Black Forest
region by the Centre for Technology Assessment in Baden Württemberg
Stuttgart, Germany (a not-for-profit think tank supported by the
state of Baden Württemberg, henceforth referred to as the 'Academy').
The purpose of the project was to reach broad citizen agreement
regarding a risk management problem, namely the siting of a municipal
waste incinerator and two biological/mechanical municipal waste
disposal plants (aerobic digesters) in the region. Plants such as these
are deemed necessary, as federal legislation has stipulated that all

waste will need to be treated (either incinerated or aerobically digested) by 2005 since the country's existing landfill sites will be prohibited from taking untreated waste. To reduce the need to transport waste over long distances to large centralized disposal facilities, the legislation also calls for a regional network of waste disposal facilities.

The project was successful; to everyone's surprise citizen panels agreed on three sites within the allotted time. Regional politicians have not yet implemented the recommendations of the panels, however, citing new information on excess capacity at waste incinerators in nearby Tubingen and Stuttgart. In this case study I look at the importance of trust throughout the citizen panel process. The role of trust and how it evolves is examined from the perspectives of the panellists themselves, the media, and local policy-makers (here the term is used for local mayors, local MPs, heads of the various city/town councils, and heads of the various ministries in the state of Baden-Württemberg).

Background

In 1992 the four local councils in the North Black Forest Region (Calw, Enzkreis, Freudenstadt and Pforzheim) formed an intra-regional corporation called PAN (*Gesellschaft zur Planung der Restabfallbehandlung in der Region Norschwarzwald*). PAN is made up of representatives of each county's Parliament, a governing board consisting of their leading executives, and a professional staff with backgrounds in engineering, economics and local planning. It was established to achieve the intraregional waste plan as stipulated in federal legislation passed by the German federal authorities (*Technische Anleitung Siedlungsabfall*). After setting the issue of waste reduction potential and the selection of appropriate waste reduction technologies, PAN's main task was to identify suitable areas in the region for a waste plant. This was put out to tender and awarded to a highly reputable engineering company, Buro Fichtner, in Stuttgart. The study was completed in 1995, and it identified no fewer than 228 possible sites, of which 11 were deemed the most suitable. Of these sites, five were considered suitable for a waste incinerator (called hot) and six for an aerobic digester (called cold). In 1994 PAN, impressed by the results of Professor Ortwin Renn's citizen panel projects in Switzerland,[19] asked him and his colleagues at the Academy to conduct a citizen participation study to help identify three sites (two cold and one hot) based on a series of criteria including suitability for transport, nearness to major municipal waste sources, and environmental considerations for further exploration.

Using random sampling techniques, Renn and his colleagues invited 5,440 citizens from the region to participate in the panels, and of these 198 accepted and 191 actually participated. The participants where divided into ten groups, of which four focused on the siting of the 'hot' plant and six on the 'cold'. Each group was facilitated by two expert moderators (either from the Academy or recruited externally). The project lasted six months, during which the participants considered written information and oral testimony from experts covering the various waste technologies and geographical information on the different sites. They also visited the eleven proposed waste sites, as well as one waste incinerator in southern Germany and an aerobic digester in northern Germany.

PAN hoped that with ordinary citizens involved in the decision-making process the final sitings of the plants would be more publicly acceptable. It was initially made clear, however, that the panels would only provide recommendations to PAN, and would have no legislative power to implement them.

The citizen panel process was beset by a range of problems. Many local policy-makers felt they were inadequately informed since they received information about the proposed sites only very late in the process (after the assessment of Büro Fichtner was completed and became public knowledge). In fact, due to a lack of communication between the various policy-maker levels (e.g., the councillors not speaking to the local mayors), several of the mayors publicly complained that they received the information only after the panellists. Unfortunately, this prejudiced some policy-makers towards the concept,[20] creating a lack of trust at the outset of the panel process. It also made it more difficult to gain political acceptance of the process once it started.

The motives of the panellists initially appear to be self-interest (to prevent the waste plants being sited near their homes or communities); they were not necessarily motivated to find the best waste management solution for the *region*. In addition, although the sample was random, only about 3 per cent volunteered to take part, raising doubts about the quality of representation itself. The lack of interest in the citizen panels by the public (although not in the waste debate as a whole) was also illustrated by the low level of attendance at the various panel processes and the waste issue exhibits arranged by the Academy in town halls throughout the region. In Horb, for example, only six citizens turned up even though the event was publicized in the local newspaper. The mayor of Horb, Michael Theurer, blamed the low turn-out on poor advance warning and non-evening hours, while the Academy felt it was more due to a lack of interest.

In an effort to avoid political manipulation of the panels, individuals involved with local and/or national politics were excluded from participating by the Academy. This further alienated local policy-makers, many of whom expressed their critical views to the local press. In areas where there were high concentrations of political activists they tried to discredit the selection process by noting it was no longer random since politicians (also citizens) were allowed to participate.[21]

The process itself was also hampered by the accuracy of the information supplied to the panels. There was a debate in the media regarding how much waste was actually produced in the region, and if it was indeed necessary to build two aerobic digesters and one waste incinerator. Hence many policy-makers argued that the assumptions on which the panellists were working were flawed. The amount of waste produced in the region is indeed decreasing significantly thanks to increased recycling. In the state of Baden-Württemberg, for example, the average citizen in 1996 produced 263 kilos of waste, only half that produced in 1991;[22] hence, many policy-makers felt the entire project was pointless, including Michael Theurer. By the year 2005, the amount of waste produced in the North Black Forest would be small enough to resolve by alternative means.[23] Yet this ignores legislation put forward by the German government: no matter how much waste is produced (and the amount of waste should not fall significantly below present levels), interregional solutions must be in place by the year 2005. Landfills are no longer allowed. In fact, some experts argue that the reason for excess capacity at various incinerators in southern Germany at present is not recycling per se, but rather the fact that landfill owners are desperate to fill their sites ahead of the 2005 deadline and are therefore undercutting the waste dumping fees of the incinerator operators.

Throughout the process, the media played a major role in reporting the views of the different actors involved in the process.

Methodology

This case study is based on two methodologies: in-depth qualitative interviews and a content analysis. Specifically, the study is based on in-depth qualitative interviews with one panellist from each of the ten panels; interviews with two of the leading policy-makers in the region (Mayor Michael Theurer of Horb, who was one of the most vocal critics to the citizen panel concept, and Mayor Sigberd Frank of Pforzheim, who was also the Chair of PAN, the group that funded the citizen panels),[24] and interviews with members of the Academy who actively participated in the project. These qualitative interviews with the two mayors and with the ten panellists lasted anywhere from 50 minutes to over two hours. The object of these interviews was to explore in-depth the panellists'/mayors' feelings towards the citizen panel project. For the interviews a detailed questionnaire was produced which was examined and scrutinized by the researchers at the Academy for both content and clarity.

The weakness of this aspect of the study is that the sample is non-representative. I only interviewed ten people for this case. In this regard, the views expressed by these ten individuals are not necessarily the same as the views of the other remaining panellists. That said, by using the in-depth qualitative interviews I was able to uncover a large amount of information which would not have been made available via a standard quantitative study.

In addition a content analysis of all available newspapers in the region (*Schwarzwalder Bote, Stuttgarter Zeitung* and *Südwestpresse*) was conducted from November 1995 (one month before the start of the project) until March 1997. The content analysis consisted of examining all the articles from these local and regional papers written on the topic. The analysis was greatly helped by the Academy's media archive. The purpose of the content analysis was both to gather background information for the qualitative questionnaire, and to gauge the media's attitudes toward the panels; editorials and 'letters' were particularly important.

The study was conducted when I was seconded to the Academy for a two-month period in early 1997. The interviews were all conducted in German by the author.

The citizen panel concept

Citizen panels (sometimes called planning cells) originate in Germany where they have been used since 1972 to give citizens a role in local planning.[25] Unlike the citizen jury process, in which the jurors are actually empowered to make decisions for a certain community, the role of the panellists is to offer advice to policy-makers. The first test run took place in 1972–3 in Schwelm, Germany, where citizens took part in the planning of a waste disposal facility. From the 1970s to the present time approximately 26 cities throughout Germany have used citizen panels as a method for local planning, and more than 2,600 adults have participated in these citizen panels to date. Citizen panels are not problem-free, however. They are not useful in helping to solve disputes where major inequities between social groups or regions are present. Additionally, as the panellists are not responsible for their actions (they only provide advice), they cannot be relied upon for accountability or long-term planning.

Results

The results of this case study are divided into several parts. Although all the data were gathered retrospectively, the survey instrument used in the interviews covered the panellists' and policy-makers' views before, during and after the process and the results follow this format with a particular reference to the role of trust. The first part focuses on the panellists', policy-makers' and media's views on the citizen panels and the actors involved (PAN and the Academy) at the outset of the process. The next part focuses on the policy-makers', panellists' and the media's views during the process itself; while the third part looks at the same groups' views toward the process once the citizen panel project was completed. The final part describes the current situation.

Views at the outset of the process (November 1995–January 1996)

With the exception of PAN, which funded the process and therefore supported the Academy, the other actors were either neutral or somewhat sceptical. A person from one panel said:[26]

> I only wanted to participate to make sure that we didn't get a disposal facility in our village. In fact, I was sure that most of the people who signed up felt the same way. They all came from

communities where a site was proposed and I guess that even though they were sceptical, maybe they felt they could do something useful for their respective villages. Who in their right mind would want a waste treatment plant next door anyway?

(Heike, Nagold)

Another participant felt that receiving money for participating was odd and thought it might be a bribe:

I almost didn't take part in the study as I felt it was very strange that I would be paid to participate (700 DM). It made me very distrustful of the whole process as it felt like a bribe and I wondered whether the Academy had already developed a solution. In the end I went along anyway to make sure the plant would not be sited in my village.

(Helmut, Eutingen)

The views of the politicians interviewed for the study differed from one another regarding the potential of the process. Mayor Frank from Pforzheim, who was also the Chair of PAN, felt that the process was important for a series of reasons:

I have always been a proponent of the citizen panel process, and I had no misgivings on awarding the Academy the contract to conduct such a process here in North Black Forest. I felt then, as I do now, that a citizen panel process in cases such as this is necessary as we must have a more open democracy in Germany. The public is becoming less and less vocal on issues, and this is changing the political landscape. It is important that the public participate and see the benefits of contributing to the policy-making process. Especially on issues such as waste management where nobody really wants these plants in their community, public involvement can provide local insights and improve people's acceptance of a need for a facility.

Mayor Theurer from Horb was more sceptical:

I am not against citizen panels per se, but I had problems with this example. Firstly, the Academy gave the impression that the concept was new and innovative and that it was useful for the democratic process and should be used here. This appeared very arrogant on their part. I did not feel that we [local policy-makers and citizens of Horb] could question the process. Secondly, Horb has undergone a significant change since I became Mayor. We have used citizen panellists and have widespread public participation in policy-making. Therefore, the citizen panel process for the waste plant sitings was not fully relevant for Horb. I mean why do we need to experiment with citizen panels here when we firstly have already had them previously, and secondly already have a strong participatory democracy in the region with high level of public involvement in local (village) policy-making – it is different in the northern part of the region, where in places like Pforzheim, the public are not involved in policy-making and feel alienated from their policy-makers.

The local newspapers were largely positive regarding the citizen panel concept in the two months preceding the project. For example, in November 1995 there were ten articles on the subject, of which only two had statements by local policy-makers criticizing it. At this stage no opinion/editorial pieces appeared against the concept. The two negative articles picked up on the issue, mentioned earlier, of politicians criticizing the process as they were excluded from the panels.

The role of local policy-makers in the panels

Mayor Theurer was concerned that local policy-makers from Horb were not included in the panels:

There are 'policy-makers' and there are 'policy-makers'. In this part of the region people are very active in politics, and as I said before we have participatory democracy here. Of 25,000 citizens approximately 200 are active in politics. This is much higher than in Pforzheim where the figure is about 50 individuals per 100,000 people. Also a large number of the local politicians do have community interests as their top priority.

The issue of inclusion of policy-makers in the process was also brought up by the panellists; Lisa from Motzingen said: 'I felt that it was not necessary to have local policy-makers involved. They wanted to maintain some control and by being cut out they weren't able to. We handled the issues perfectly well without them. I really felt that the issue was exaggerated out of all proportion.' Another panellist, Johan from Horb, said: 'I do feel that Theurer had a point about local policy-makers having a right to participate as they are citizens too, but the panel process was not affected by not having them participating.'

The Academy held several views on this. Dr Sabine Körbele, who was responsible for the political context in the citizen juries, felt that:

> In retrospect maybe we should have included them. The trouble that some policy-makers made about their exclusion did affect the credibility of the process early on, and it could have been avoided. I don't feel that having some policy-makers on the panels would have influenced the outcome significantly. However, that said, as policy-makers have other ways to influence the process, I still believe in principle that it was correct not to include them.

Yet Professor Renn, one of the Directors of the Academy, felt that they were right to stand by the original decision since the object of the panels was to seek the participation of the public and not the policy-makers who had other channels to voice their views.[27]

Summary of the views at the outset of the process

At the outset of the process views were mixed. The citizen panellists interviewed were somewhat sceptical about the process and mostly participated to ensure the waste plants would not be sited in their neighbourhoods while the media was largely neutral. As expected, of the policy-makers interviewed, Frank was largely positive and Theurer largely negative.

The role of trust

Already at this stage trust was an important variable. Citizens participated because they did not trust the siting process, and they wished to prevent the waste plant coming to their neighbourhood. Local policy-makers meanwhile tried to discredit the process by questioning its arbitrariness. Feeling excluded, they believed they could inject a feeling of distrust among the public. For example, one of the panellists interviewed mentioned that on several occasions policy-makers had mentioned in passing that the whole process was needed as a strategy was already in place to deal with the waste issue.

The Academy, however, stood by its assertion that including policy-makers would exert undue influence on the process. Looking at the actual numbers of participants this would seem to be a correct. Since few of the random sample of the public actually agreed to participate (200 out of 5,440 citizens, or approximately 3 per cent), interested policy-makers would probably have carried much greater proportional weight if they were included.

The citizen panels in action (January–June 1996)

Once the project was under way, criticism from the excluded policy-makers intensified. In Horb they went as far as hiring a consultant to advise them on the citizen panel process itself. Although they had conducted one previously, they still felt they lacked the necessary expertise. This consultant was also employed to contribute to news-papers articles and participate in media and council discussions on the 'problems with the process', fuelling mistrust rather than minimizing it.

Yet over time criticism subsided among the policy-makers. Before the publication of the panel findings, the media began to argue in favour of the process. The panellists themselves were surprised about how their points of view changed over time and their own scepticism subsided. One panellist said:

> At the first meeting I was highly sceptical. I did not believe that we would come to any firm conclusion. I also did not like Prof. Renn; I felt that he was very distant from the panellists. However, by the end of the process I was quite happy about the whole thing. We had worked hard and came to an agreement with the other panellists which was satisfactory. I even felt that I could trust Prof. Renn.
>
> (Wolfgang, Pforzheim)

Another panellist described how she changed her mind:

> I joined the process as I did not want the waste plant in my village. I mean no one can trust the policy-makers to help you come up with a right decision, so it is better to be a part of it. I am happy with the solution that we came up with and I think that it

was the right one. However, retrospectively, I do feel that the site close to my village would also have been appropriate. And I wouldn't have minded so much as it was better suited than several others in the region.

(Heike, Nagold)

The policy-makers interviewed did not change their views significantly throughout the process. Frank was positive throughout:

I have supported the process from beginning to end. I was sure that Prof. Renn and his colleagues would do a good job and they did. I really did not change my mind during the citizen panel process. The outcome is more than satisfactory. However, many of the policy-makers in the region did not want an aerobic digester or an incinerator in their community and tried to discredit the process. They were narrow minded and ignored the big picture. Although some of these views can be understood as local policy-makers are concerned about their voters, I was surprised that the Greens also acted in this way. They say they are open-minded about different solutions, but in actuality this was not the case.

Mayor Theurer did change his view of the process a little over time. He denied that he tried to discredit the process but felt that it was important to have an external consultant review it:

I am not an expert on citizen panels and neither are my colleagues. Hence, I felt it was necessary to buy in this expertise and that is why I hired a consultant. He provided the expertise that we did not have and he showed that a lot of what the Academy was saying was incorrect.

['Like what?']

Like the fact that the panel idea was new to the region. Also, you must not forget, there was a great deal of mistrust generated against the Academy. A large amount of this distrust was removed during the citizen panel process although personally, I still am distrustful of some members in the Academy.

He reiterated his concern that local policy-makers should have been allowed to participate: 'I remain unhappy that local policy-makers were not able to participate in the panels but I think that the concept is useful and important to try so I tried to be less critical.'

The role of the media

The content analysis also shows a more positive attitude developing over the time they were operational. In January 1996, at the outset of the process, there were 14 articles criticizing the panels out of a total of 30 (only 2 were positive while 14 were neutral), but in May only 6 out of 29 articles were negative (21 were neutral and 2 positive). The initial articles against the process were scathing in their criticism, comparing the panels to a game show without clear results masquerading as democracy, and branding the Academy as nothing more than a well-paid accomplice.[28] At the first panel sessions panellists were quoted as saying, 'We wasted four hours' and 'The process did not impress me', implying that the process was manipulated by the mediators.[29] Another editorial criticized the randomness of the process as the citizens themselves could choose in which group they participated. As a result, people from the same town/region could participate in the same panel.[30] In the first two months politicians criticized the process as scandalous; only 200 people could participate and was therefore hardly 'participatory democracy'.[31]

Further discussion focused on the policy-makers' concern that the public did not have enough knowledge. This was well illustrated in a heated debate in Horb where Renn and one of his workers, Dr Körbele, argued against Mayor Theurer regarding the wisdom of the public. Körbele asked whether the Mayor believed members of the public were as stupid as cows, a rebuke which the policy-makers did not properly address. Renn added that one should never underestimate the knowledge of local citizens.

As those policy-makers who were critical of the process were unsuccessful in killing it off, they adopted a different tactic: to provide as much information as possible to the citizen panellists to help them make the right decision. This was commented on by one of the panellists:

> The Mayor in Horb organized several information meetings for the panellists in the Horb area where he himself participated. He wanted to make sure that we had the right information at hand. I really felt that Mayor Theurer acted most professionally and I have high respect for him.
>
> (Johann, Horb)

Another panellist said, however:

> I can't believe how much information we received. I felt that we were over informed and I could not read all of it. It was simply far too much. It would have been better if we had got less information and the information we received had been better targeted.
>
> (Sigrid, Pforzheim)

As this process continued, policy-makers could no longer openly criticize it since that would bring into question their commitment to democracy and the empowerment of the people. As their criticisms subsided so did the media's, and simultaneously the panellists' confidence grew as they were making progress on where to site the waste plants.

While criticism of the process could be ascribed to a lack of understanding and unfamiliarity with the concept in these early stages, the hiring of a consultant in Horb appeared to directly undermine the process by attacking the Academy's integrity.[32] Renn and his colleagues commented on some of these claims, reiterating that active citizens could participate in the policy-making process but not politicians; and that what the consultant (Friedrich) was saying was untrue.[33]

As the panels grew more cohesive, criticisms on all sides declined. One panellist said: 'In the beginning I was unsure what I was doing there, but we grew into a team after a while. It was like it was us against the various pressure groups and disgruntled local policy-makers and we wanted to make the right decision.' Another stated:

> The claims of us being manipulated really got on my nerves and at first I did wonder about the Academy's motives, but as time went on and as I got on with my work, I felt that the Academy were independent and did the job very well, and we became a team.
>
> (Joseph, Neuligen)

Even the media became more positive: 'Cool. The whole process is going really well... The citizens have made a great step in the right direction.'[34] The citizen panellists were even prepared to defend themselves in the press, explaining why they had decided on the particular locations for the waste facilities.

Summary of the citizen panels in action

When the project actually began, the campaign by some local policy-makers to discredit the process continued and in some aspects intensified (such as the hiring of a consultant in Horb). The main objective appears to have been to pressure the Baden-Württemberg government into cancelling the project. In other words, by injecting distrust into the process it would be discredited by the public and regional policy-makers.

The role of trust in the action phase

It has been shown that the panellists themselves did not trust the process and that is why they participated in the exercise. By the end of the second part, the citizens came to believe in the project, even considering a waste plant in their village if they saw the location as suitable.

One of the main reasons why trust was implanted among the citizen panellists was the perceived competence of the Academy and the mediators. This was not an easy or straightforward task, however. Local policy-makers, particularly in Horb, tried to discredit the process if they could not also participate. This only backfired; the arguments they and their hired consultant devised were proved to be unfounded by the Academy, in whom the panellists now put their trust. According to one of these: 'Policy-makers care only about themselves. They are driven by power and greed and are little concerned about the public except during elections. How can you trust them? They will say one thing one day and another the next.' Another panellist said: 'This is all a game. Politicians are in the hands of industry. Industry has power and money and this is what the policy-makers want. We have no money and no power so why should they care about us?' Yet another panellist focused on how the panellists were influenced by the local policy-makers: 'They complained quite a lot, but we just got on with it. I don't trust local policy-makers as they are simply politicians, and this episode did not make me trust them any more.'

Outcome and directly afterwards (June–November 1996)

The press and most of the policy-makers were pleased with the outcome. Against all expectations the panels agreed that one aerobic digester should be built in the south of the region (they felt that no particular site in the south was suitable but Horb was the highest on the shortlist) and that the other aerobic digester and the incinerator should be built in the north of the region (the incinerator in Pforzheim and the other aerobic digester near Pforzheim). These were the views expressed by some of the panellists: 'When we finished the process I was extremely happy. We had done a great deal of work. I couldn't believe that it ever would be finished. The solutions proposed were the right ones.' Some felt that the process would be useful in other regions: 'Of course I would recommend the concept to other regions. It worked well here and I am rather pleased.' Others, although generally happy, felt that improvements could be made: 'It is a concept that definitely should be repeated. But I didn't like the debate surrounding whether politicians should be allowed to participate in the process or not. Next time this happens, let's put the local politicians in a separate group and make them happy!' Finally, a citizen panellist complained about the lack of time: 'I liked the whole idea even though we had too little time. It is a way for local citizens to participate in democracy and therefore it is something that should be repeated.'

Policy-makers were generally happy with the outcome of the process when the results were announced. Frank said, for example:

I am especially pleased that the citizen panels were able to agree on three sites for the waste plants. It shows that the process worked. You know this issue that the citizens were not smart enough and therefore needed help from policy-makers was completely unfounded. I mean the citizen panellists know 90 per cent more about the handling of wastes than policy-makers, and as you see I think the decision that they made was the right one.

Even Theurer was almost convinced: 'In the end the panellists did not do a bad job, but I still wonder what the outcome will be. I wouldn't be surprised if Horb was selected in the end anyway.'

Analysis of the risk factors

Examining the case studies in light of the risk factors outlined in Chapter 1, one can draw the conclusions as follows.

In a high public trust high/low uncertainty risk situation, deliberative risk management strategies are not required. This factor does not apply to this case study, where there was high public distrust.

In a low public trust situation, some form of risk management strategy will need to be implemented, but this depends fundamentally upon the reasons for this distrust in the first place. As stated in Chapter 1, there are three reasons why the public does not trust regulators: lack of impartiality, incompetence, or inefficiency. In this case, only one of the factors is relevant. The public did not believe the local regulators to be impartial. The regulators were obliged to put forward an intra-regional solution for the waste problem by the German government. They were thus acting on behalf of the state, rather than the local public. As this was the case, a deliberative process was called for.

Competence was not seen as an issue. Indeed, the process up until the involvement of the publics in the citizen panels was conducted with the utmost competence. The Stuttgart-based engineering bureau (Fichtner) identified over 200 suitable sites for the waste incinerator and the two aerobic digesters. The issue of competence was only raised after the implementation of the deliberative approach, and this was an attempt by local policy-makers opposed to the approach to undermine it by injecting an element of mistrust. Inefficiency was never an issue either. The process up to the involvement of the citizen panels was seen as efficient, and neither local policy-makers nor the public suggested a misuse of public funds.

Deliberative techniques can help create public trust regarding a contentious risk management issue – if the public mistrust issue has something to do with partiality – but these techniques are expensive and time-consuming. In this case, public mistrust was based on a conception of impartiality. Through the implementation of the deliberative approach, public trust was created in the regulatory process. Indeed, the citizen panel approach in this case study was hugely successful; the panellists' agreed final outcome was supported by all the actors involved, including an initially hostile media.

The deliberative approach, however, did come at a price. It was both expensive (it cost approximately $1 million) and time-consuming. The citizens had six months to come up with a decision for siting positions for the waste incinerator and the two aerobic digesters.

In high distrust situations, charismatic individuals are extremely helpful in negotiating successful deliberative outcomes. Although there was high

public distrust in this Black Forest case, there were no real charismatic individuals. Trust was still built up in the risk management
process, however, through a high level of competence among the
Academy mediation team led by Ortwin Renn, and the citizen panellists themselves who withstood criticisms from local policy-makers
and the media.

*In any regulatory/risk management process, local or national political
actors have to support the final outcome.* This did not occur in this
case. Mayor Frank of Pforzheim, arguably, the most powerful actor
in the area and also from the largest town in the area, who had put
so much effort into getting the citizen panels off the ground in the
first place, opposed the final outcome. He did not want the incinerator
in 'his' town. Due to his opposition, the consensus that was
developed through the citizen panel process unravelled. This led to
greater public distrust, where in effect some publics opposed other
publics, than had the deliberative process not been conducted in
the first place.

The regulator cannot assume public trust, nor take it for granted. In this
case, PAN presumed a lack of trust existed. As a result Mayor Frank
favoured a deliberative approach, though PAN did not actually test
for public trust. Results of such a survey could have better pin-
pointed why the public distrusted PAN in the first place. In so doing
the deliberative process could possibly have been better targeted and
developed.

Proactive regulation is more likely to gain public trust. The Black Forest
case is an example of proactive regulation in which citizens themselves could help select where to put the waste incinerator and the two
aerobic digesters. In so doing a policy vacuum never occurred. PAN
and some local policy-makers decided to do this; they assumed the
public was mistrustful since there is a history of opposition to the
siting and building of waste incinerators in Germany and elsewhere.

*Interest groups will in many cases try to create public distrust of regulators
which in turn can lead to failures of the risk management process.* In this
case interest groups did not try to discredit the regulators per se, but
rather the process of deliberative risk management process as a whole.
They too wanted to have an active role in determining where to site
and build the three waste plants; citizen panels diminished this.
NGOS, citizen activists and other special interest groups are used to
'representing the public'. When this role was fulfilled instead by the

citizen panels, they questioned their expediency. But the mediator, Ortwin Renn, the Academy, and the citizen panellists themselves enjoyed a higher level of credibility in this case than the special interest groups.

Interest groups are needed, however, when the regulator is not seen as impartial and when one is dealing with national or international regulatory issues. This risk management factor does not apply to the German case, since it was local.

4
Risk Management in the United States: The Case of International Paper's Hydro-Dam Re-Licensing Procedure

Introduction

If we examine the ideal types summarized in Chapter 2, the USA case stands out. It encompasses all four components in varying degrees. The regulatory regime used more openly in the USA than other countries surveyed in this book is a rational risk policy on strict economic grounds. This, highlighted by the OMB's active involvement in regulatory policy-making in the USA, was an approach first made popular in the Nixon and Ford administrations. Cost-benefit analysis, cost-life analysis, and so on are therefore frequently invoked in the policy-making process. The USA also has a technocratic/expert element branch in regulation. An example of this is the EPA's Science Advisory Board, which is frequently asked to comment on proposed regulations.[1] The US regulatory system also has a well-advanced deliberative component. Initially enshrined in legislation (e.g., the National Environmental Policy Act of 1970) which actively encourages public and interest group participation in the policy-making process, it has more recently embraced negotiated rule-making, made law in the 1990 Negotiated Rulemaking Act.

The US approach to risk management can be termed 'adversarial', as the process is inherently litigious and expensive and much more common in the USA than elsewhere.[2] That said, Robert Kagan argues: 'Legal contestation in the United States is so cumbersome, costly, and frightening that disputants resolve most conflicts and litigated disputes by informal negotiation.'[3] The process is also unusually transparent, however, and thus open to a wide range of stakeholders and the public. Sheila Jasanoff characterizes the US risk decision-making process as 'costly, confrontational, litigious, formal and unusually open to participation'.[4]

History of the adversarial style of regulation

The adversarial approach comes from England, where it was made popular during the opposition to royal absolutism around the Civil War in the 1640s. The philosophical origins of the perspective, in which arguably everybody can participate, can be found in the work of Hobbes, Locke, Moore and Plato who all argued that at the starting point of civilization there was no political community, just individuals.[5]

The present process of regulation in the USA was taken from England by its emigrants. The US Founding Fathers saw it as a departure from – and improvement on – the centralist system and monarchy in England, particularly under the rule of King George III. They called for a dual 'State'-level (local) and 'Federal' (national) system of government with an elaborate checks and balances approach both to prevent political absolutism and to ensure competition between the different parts of government: an Executive Branch (including the President and its offices[6]), a Legislative Branch (comprised of a Senate and a House of Representatives), and a Judicial Branch. These bodies, as well as the individual States of the American Union, together develop and pass laws and regulations. Through this process they are in constant battle with each other for power and influence.[7] Unlike the other countries surveyed in this book, which sought consensus, compromise and collaboration (albeit not before 1842 and the passage of the Pollution Control laws which encouraged collaboration between industry and government) so as to build trust between the various players in the regulatory process, the process in the USA is based on distrust of autocracy.

US regulation since 1945

From the end of the Second World War until the mid-1960s, the outcomes of the US style of regulation were not significantly different from those in Europe.[8] During this period the American public was generally supportive of business, believing that a strong industrial sector was essential in meeting the Soviet threat and for the USA to continue to assert its authority in the world.[9] Washington trusted industry to manage its own affairs, a decision reinforced by a legacy of remarkable growth and expansion. Legitimacy was never an issue at this stage: regulators were seen as experts who simply tried to attain certain goals (e.g., a cleaner environment and a safer work place).[10]

The state of US regulation changed drastically after a dramatic growth in worldwide public awareness of environmental and public health issues following groundbreaking books such as Rachel Carson's *Silent Spring* in 1962 and the Club of Rome's Limits to Growth study in 1972, as well as highly publicized environmental disasters including Minimata in Japan and Seveso in Italy. Indeed between 1965 and 1975, according to Vogel, 'more legislation was enacted and more new regulatory agencies were established to administer them than in the entire history of the federal government'.[11]

Following this shift in worldwide opinion on environmental and health issues, US politicians quickly capitalized on the issue. Edward Muskie, for example, who was the vice-presidential candidate in the 1968 election and was widely tipped as the Democratic nominee to take on Nixon in the 1972 election, pushed environmental issues very hard.[12] Indeed, during 1970 there were a series of political events that could be seen as part of the 'environmental bandwagon'. First, the EPA was founded, legislation was passed to combat air pollution (the Clean Air Amendments) and the National Environmental Protection Act (NEPA) came into being, formulating general environmental policy. At the time, US environmental groups welcomed the formation of the EPA, as they were convinced that an independent, mission-driven agency was needed to avoid capture by industrial interests. The establishment of the EPA led to a shift in policy-making authority from Congress to administrative agencies, leading to a change from the system of 'shared process' to shared influence over bureaucratic decision-making.[13]

Throughout this period, the public increased the pressure on regulators via elected policy-makers. Driven by regulator concern and media horror stories, the public demanded strict regulation of industry, in particular large corporations, which were seen as the arrogant elite. Industry, rightly or wrongly, was blamed for the environmental damage that the media and environmental groups were amplifying.

Under the newly formed EPA, led by William Ruckelshaus, the former Assistant Attorney General of the state of Indiana, lawyers acting on behalf of the Agency became involved in dealing with pollution regulation. In the EPA's first two months of existence, Ruckelshaus brought five times as many enforcement actions as the agencies he had inherited had during a similar period, partially so as to demonstrate good faith and commitment to environmental objectives.[14] As John Quarels, his General Counsel, stated:

> Ruckelshaus believed in the strength of public opinion and public support...He did not seek support for his actions in the established structures of political power. He turned instead directly to the press and public opinion...The results were impressive, especially during the period of public clamor for environmental reform.[15]

Similarly, on workers' safety, the newly established Occupational Safety and Health Administration (OSHA) pushed for strict controls on industry, levying fines even for small violations.[16] Studies showed that these regulations, using new tools and techniques (particularly the threat of legal action), were successful in ensuring that regulators avoided the accusation of regulatory capture and in so doing kept their legislative mandates.[17] These command and control guidelines passed by Congress also made agencies more accountable as a whole and industry became less influential in making regulations in the process.[18]

In addition, the public itself became more actively involved in the policy-making process. The 1970 Clean Air Act gave the public the right to sue regulatory agencies so as to trigger policy enforcement.[19] The aspect of suing became more prominent following the passage both of the revamped 1993 Administrative Procedure Act, which acts as a notice and comment procedure before a regulation is made into law, and the 1996 Administrative Dispute Resolution Act, which stipulates that consultation with a range of stakeholders must take place before an agency issues a proposed regulation.[20]

These measures put a heavy strain on the country's industries. They felt unfairly treated and began taking legal measures to protect their interests so as to be able to adjust to a dramatically changed regulatory environment. Their reaction was understandable. They were increasingly being sued by the EPA and the public as well as by interest groups via 'citizen suits'; at the same time the EPA, taking the view that it would enforce the country's pollution control laws, levied more and more fines on them. Industry regulation costs increased from $147 million in 1975 to $268 million in 1977, a jump of 82 per cent. As a result, industry decided to start fighting the special interest groups and regulators by taking them to court.[21] Threats of litigation could also delay decisions for years, working in favour of industry which on the whole had more money to spend on lawyers than the regulators.[22] The problems were compounded by the eager yet inexperienced regulatory bodies which began making mistakes. Several issues were handled poorly, and the EPA suffered credibility problems.[23] In the face of this backlash from industry the public demanded even stricter regulation but policy-makers were hampered by the national 'strategic' costs of such legislation on industry in the wake of the 1973 oil crisis. Regulators therefore decided to push for a more restrained response.[24]

By the 1980s, however, these regulators were under renewed and increased pressure from Congress and its General Accounting Office (GAO). In addition, advocates for the tighter environmental and health protection standards became dismayed by the regulators' repeated failure to meet deadlines and to respond forcefully to recognized hazards. They were particularly upset at the limited success of the EPA, realizing belatedly that the Agency's staff were relatively inexperienced.[25]

In response to this criticism the EPA began quantitative risk assessments to justify their regulatory decisions and to quantify uncertainties as much as possible. In effect the EPA took the view that 'numbers' (particularly if they were shown in a transparent fashion) could be trusted more than experts' (be they in-house or independent scientists) qualitative judgements (this was the reverse of the arguments in the UK).[26] Quantitative risk assessment measures were further popularized following Reagan's and Clinton's guidelines for strict cost-benefit analysis to ensure regulations would not become overly expensive for industry to implement.[27] This extra layer of analysis, however, only made the regulatory process more complex, slower, and less effective.

The battle for regulatory authority

The Administrative Procedures Act of 1946, which re-established the authority of the courts and lawyers in the regulatory process, was challenged by a wide array of bodies in the 1970s. Most importantly, in 1974 President Ford began to rein in the power of the agencies and the courts by authorizing the OMB to assess the inflationary impact of proposed rules. Executive control was further tightened during the Reagan Administration, with an Executive-based policy of regulatory relief through OMB. This forced regulators to provide detailed cost-benefit analysis for its proposed regulations.[28] This requirement for regulatory agencies to submit all proposed rules to OMB for pre-publication review – which would have an effect of more than $100 million on the economy – is still in place today. To expedite this, agencies such as the EPA developed a capacity to analyse these regulations via their own in-house cost benefit analyses.[29] Hence, in the policy arena, conflict was then continuously generated,[30] with the agency on one side, the OMB on the other and somewhere in the middle citizen groups, NGOs and industry.

Regulatory reform

The risk management process in the USA is continuously tinkered with. Policy-makers know that there are significant problems with the current use of command and control regulation,[31] and the problems can be summarized as follows:

- uniform standards are economically inefficient
- regulations are not inherently based on a proper cost-benefit analysis
- development and revision of standards is slow[32]
- end-of-pipe solutions and not pollution prevention are encouraged[33]
- incentives for firms to go beyond compliance are not provided
- the regulatory process itself is unusually adversarial and legalistic[34]
- the process is too fragmented with media-specific, pollutant-specific and sector-specific approaches.[35]

In the making of regulatory reform, there are presently three trends in US risk management thinking, which roughly correspond with the three risk management strategies outlined in Chapter 2: rational economic risk policy, risk–risk ranking (technocracy), and striving for a consensus via negotiated rule-making (deliberation). Rational economic risk policy and risk–risk ranking exercises are discussed in Chapter 2. Negotiated rule-making, the subject of the US case study, is briefly outlined below.

Negotiated rule-making: the dams on the Androscoggin river

There has been a strong push to use deliberation to increase consensus in the policy-making process. For example, for a ten-year period (until 1993) the US EPA was actively promoting such approaches as it recognized that high litigation costs had to be reduced. It asserted (without any analysis) that nearly 80 per cent of the 300 regulations that it put forward each year ended up in court.[36] Although this figure has been disputed (most notably by Coglianese, who argues that the number is actually 30 per cent),[37] the agency took the view that negotiated rule-making would save time and money,[38] as the traditional regulatory process (in particular with regard to permits and licensing) was not only rule-bound but highly adversarial.[39] The regulator has been unwilling to offer guidance to the regulatee on how it could conform to the standards. Thomas Kelly, Director of EPA's Office of Standards and Regulations, argued at congressional hearings in 1988 that negotiated rule-making would help EPA avoid a 'regulate, litigate, regulate, litigate syndrome'.[40]

The idea of negotiated rule-making in the USA goes back to the early part of the twentieth century, when it was used by the Federal Trade Commission.[41] The need to include a range of interested parties in negotiations came up again around the time of the New Deal in the 1930s,[42] but it was not until the mid-1970s (under the then Secretary of Labor, John Dunlop) that negotiated rule-making came back to the fore. He argued that groups affected by a particular regulation should be allowed to participate in its design.[43]

The modern pioneer of negotiated rule-making is Philip Harter, who argued in his seminal 1982 paper that it was a cure for the modern regulatory malaise.[44] Around the time the paper was published, several regulatory agencies examined the potential for negotiated rule-making. The most enthusiastic agency was the EPA, which as early as 1980 showed support for the idea. In 1983 the EPA, via the *Federal Register* (a rulebook where proposed federal legislations are announced), expressed an interest in pursuing negotiated rule-making, and shortly thereafter began soliciting interest from environmental groups and industrial bodies. Consultants (ERM-McGlennon) were hired to assist in the communication process, and the Negotiation Program at the Harvard Law School was asked to comment on and evaluate the early stages of the process. This evaluation showed that negotiated rule-making had significant promise.[45]

By 1990, five federal agencies had set up guidelines for negotiated rule-making, and in the same year the US government passed the Negotiated Rule-making Act.[46] At this time, Senate and Congress were supportive of the procedure, seeing it as a way to reduce the spiralling costs of litigation.[47] Al Gore's National Performance Review suggested that negotiated rule-making was an important tool that all agencies should consider adopting.[48] In the same year, President Clinton's Executive Order 12,866 argued that 'each agency... is directed to explore, and where appropriate, use consensual mechanisms for developing regulations, including negotiated rulemaking'.[49] Since 1990 the EPA, OSHA and FERC (Federal Energy Regulatory Commission – the subject of this case study) have implemented a series of negotiated rule-making exercises.

There have been some attempts to review the outcomes of the negotiated rule-making processes that have been undertaken to date. The findings of these evaluations are mixed and can, for the sake of argument, be divided between those who believe that negotiated rule-making has some promise and those who do not. Langbein and Kerwin, for example, interviewed 50 participants in six EPA conventional and regulatory-negotiation rule-making exercises and reached the following conclusions:[50]

1   Participants in the negotiated rule-making process express greater satisfaction with the final rule than those in the conventional rule-making process.
2   Participants learn more in the negotiated rule-making process than in the conventional rule-making process.
3   Participants view negotiated rule-making as an inclusive yet resource-intensive process in which they all learned a great deal. Indeed, 78 per cent of the participants felt that the benefits outweighed the costs.
4   Data suggests that the negotiated rule-making participants are significantly more likely than conventional rule-making participants to report that the other parties will comply with the final agreed-upon rule. In sum, negotiated rule-making is likely to improve compliance.

5 Through face-to-face communication, something that negotiated rule-making encourages, the rule-making procedures help to improve social outcomes, including core relationships of trust and reputation. This latter point is consistent with the literature which shows that face-to-face communication increases the likelihood of the process being seen as fair and trustworthy.[51]

In addition, EPA's own internal evaluation of the first seven negotiated rule-making exercises point out that they produce rules more quickly and use fewer resources than conventional ones.[52]

That said, other evaluations of the process, by Caldart and Ashford, Coglianese, Freeman, Rose-Ackerman, Siegler, and Susskind and Secunda, are rather pessimistic. These researchers make the following points:

1 Negotiated rule-making has not caught on; less than 0.1 per cent of all rules have been based on negotiated rule-making (in the time period 1983–96).[53]
2 It is not a popular way of doing things among EPA's own staff, who prefer conventional regulation. There is deep scepticism within the agency about working with industry.[54]
3 Negotiated regulatory processes are driven from the top by politically appointed individuals rather than by the inspectors working on the ground, which has caused internal conflicts.[55]
4 The switch from traditionally rewarding successful enforcement actions to rewarding successful regulatory outcomes is not welcomed by enforcers.[56]
5 The negotiated rule-making process is resource-intensive and takes a considerable amount of EPA staff time.[57]
6 Negotiated rule-making does not decrease the amount of litigation so supposedly apparent in conventional forms of regulation.[58]
7 Negotiated rule-making may in fact increase conflict, through excluding some groups from participation in preference to others.[59]
8 EPA's enforcement structure and culture is a barrier to the implementation of regulatory negotiation developments.[60]
9 Negotiated rule-making cannot succeed unless all the participants have a clear idea of the actions that the instigator of the process will take if an agreement cannot be made.[61]

10 Some environmental groups are reluctant to work with government and industry on confrontational issues (such as permits) as they feel that participating may lend credibility to an unsatisfactory outcome.[62]

11 The negotiated rule-making process is in many cases unfair for industry participants *vis-à-vis* environmental NGOs, as the former frequently have to check back with their constituents regarding proposed policies being negotiated.[63]

12 Some, particularly smaller, local public interest groups do not have the capacity to participate in regulatory negotiation processes.[64]

13 Negotiated rule-making procedures are not democratically legitimate unless all interested parties can participate, which is not the case for most of them.[65]

14 Negotiated rule-making procedures hamper creativity and lead to weaker regulations overall, as the drive for consensus (among a wide array of different actors representing different interests) ensures that a weaker form of rule will be more likely to achieve unanimous agreement than a tougher one.[66]

Some of these claims have been refuted by the proponents of negotiated rule-making, most notably Harter.[67] In a recent article, Harter disputes the claim that negotiated rule-making is more time-consuming than conventional forms of rule-making, and argues that it is both time-efficient and viewed as highly positive by those who participate in the process, as well as by the agencies themselves.[68] Harter states, for example, that EPA's experience with negotiated rule-making has reduced the rule-making period by an average of a whole year. He takes the view that Coglianese's 1997 findings are misleading, pointing out, for example, that Coglianese counts negotiated rule-making processes that have in effect been abandoned into his overall calculations. As a result, EPA negotiated rule-making exercises perform considerably better than Coglianese suggests. Harter also questions Coglianese's criteria for when a negotiated rule-making procedure should be viewed as complete. With regard to the Coast Guard's negotiated rule-making concerning Vessel Response Plans, for example, Harter takes the view that the goal of the agency instigating the procedure is not the issuance of a final rule. These and other claims are vehemently dismissed by Coglianese in a reply to Harter.[69] There Coglianese states, among other findings, that the case which Harter refers to (farm worker protection) was in fact not abandoned, but rather an example of where one of the participating parties walked away from the negotiating table. In addition, through new data, Coglianese dismisses the other cases that Harter puts forward and states categorically that negotiated rule-making procedures do take more time than conventional ones and thus concludes that Harter's defence of the negotiated rule-making tool is a form of advocacy rather than a robust analysis.

Taking into account the earlier positive and negative findings associated with the use of the negotiated rule-making procedure, this chapter evaluates the use of the procedure as applied by FERC to a dam re-licensing case, by focusing on three criteria raised by other evaluators, namely cost, time and trust.

Re-licensing of hydropower dams

This case study focuses on International Paper's/Otis Power's attempts to re-license four of its hydropower stations on the Androscoggin River in central Maine. Like other pulp and paper companies in Maine, most notably Bowater/Great Northern Paper, International Paper (IP) and its subsidiary, Otis Power, owns and operates several hydropower stations. These stations generate electricity that is used in part to run the company's pulp and paper mill at Jay, Maine. As these hydropower stations are located on rivers owned by the Federal Government of the United States, licences are needed to operate them.[70] The licences for the operation of these dams have to be renewed, usually every 35–50 years. This involves the dams' owners making a case to the Federal Government that continued operation will not adversely impact the river ecosystem. This is usually done through an Environmental Impact Assessment or a Draft Environmental Assessment, backed up with a number of environmental and historical studies. Often the licence renewal process involves the licensee making some environmental concessions to ensure that the benefits of having the dam (renewable, relatively pollution-free electricity) do not exceed the environmental costs associated with it (e.g., negative effects on fish species or siltation).[71]

Every year a large number of dams come up for re-licensing. Since 1993, for example, the licences for over 260 dams across the USA have expired and over 550 more will expire in the next 15 years. Past studies show that re-licensing can be a complex regulatory affair.

Detailed analysis and a critique of the applicant's environmental impact assessment can come from the regulators and interest groups and, in addition, the ruling can be appealed and counter-appealed, resulting in the licence being challenged in the courts for up to a decade, if not longer. In 1988, for example, Bowater/Great Northern Paper Company began re-licensing procedures for two of its hydro-electric dams (out of a total of 16 it owns) along the Penobscot River in Maine. To date the issue has not been resolved. After a 10 year effort, and at a cost of approximately $11 million to the company (mainly for lawyers' fees but also for a significant number of environmental improvements), the company did receive a licence renewal to continue operating the two dams.[72] This, however, has been appealed by interest groups on environmental grounds.[73]

In other cases the re-licensing procedure has not been successful. In 1998, for example, the FERC ordered the Edwards Dam in Augusta, Maine, to be dismantled, arguing that the environmental damage caused by the dam exceeded the economic benefits. In April 1998, the same commission rejected a proposal by Bangor Hydro to build a new dam on the Penobscot River, stating that this would hinder attempts to restore wild salmon there.

In recent years, recognizing that the re-licensing procedure can be tedious, expensive and highly adversarial (all trademarks of the US regulatory system), the government, industry and other stakeholders have felt that a new approach should be adopted. The approach that regulators favoured and which a large number of interest groups agreed to try out was negotiated rule-making.[74] In fact, the FERC so favoured this new approach that it added a series of incentives to encourage applicants to consider it:

1 FERC allowed the applicant to conduct Draft Environmental Assessments rather than the more expensive and more tedious Environmental Impact Assessments.
2 The Commission would provide feedback throughout the re-licensing process. This feedback included commenting on scoping reports, examining Draft Environmental Assessments and taking part in public fora.
3 A fast-track application process. The Commission would ensure that the re-licensing procedure would not be subject to unwarranted delays.

Regulatory negotiation within the hydropower sector

Within the hydropower sector the regulatory negotiation process was first announced by a Notice of Intent in the *Federal Register* on 3 December 1996, and made law in October 1997.[75] All the interested stakeholders could be involved early on in the discussion of the re-licensing process for hydropower stations for the first time.[76] Preliminary analysis of this new collaborative re-licensing process projected that cost savings of 20–50 per cent could be realized.[77] To date, several power companies including Niagara Mohawk Power Company and the New England Power Company, as well as other industries with hydropower assets, have successfully used the regulatory negotiation procedure to reduce the cost and acrimony related to re-licensing existing dams.[78] IP's re-licensing application for its four dams on the Androscoggin River was seen as the first successful example of this new re-licensing process for a hydropower dam in USA, beginning even before the regulatory negotiation approach was codified in the FERC regulations.

Negotiated rule-making and the Androscoggin case

International Paper, and its subsidiary Otis Power, currently operate four dams along the Androscoggin River, namely the Jay, Livermore, Otis and Riley dams in west-central Maine close to International Paper's Androscoggin Mill in the town of Jay located 60 miles inland.[79] These dams operate in a run-of-river capacity in which there is a continuous flow of water through the turbines with a minimum flow of 1,245 cubic feet downstream from the dam project.[80] These four hydropower dams currently have an installed capacity of 29MW, and generate on average 111,828 MWh per year, which provides approximately 13 per cent of the energy requirements of the Androscoggin Mill. The licences for the generation of electricity at the dams were due to expire in September 1999 and, as International Paper felt that their hydropower stations generated electricity more cheaply and with less environmental impact than comparable fossil fuel sources, it sought to have them re-licensed. In fact, research on the economic benefits of the four units indicated that they generated a net benefit of $660,000 per year for International Paper.[81]

The Androscoggin Mill has had a mixed environmental and public relations record. During the 1970s and 1980s the mill was considered to be one of the most environmentally polluting in the country. By 1990 the situation had changed for the better, following a series of environmental improvements. Today the Mill has one of the best environmental records of IP's mills and is by far the most profitable. With regard to its reputation, in 1991, following a prolonged strike, IP fired the entire unionized workforce at Androscoggin, replacing them with non-unionized workers. This caused a massive outcry in Jay, the town closest to the mill, where IP was the largest employer. In fact, as an outcome of the massive lay-offs, the town of Jay adopted its Environmental Ordinance and began to pursue IP aggressively through a Code Enforcement Officer, the only such code enforcer in the USA.[82] There was massive public distrust of IP's Androscoggin operations, not only regarding pollution but also labour laws. This led to concern within IP that the re-licensing procedure could become difficult, even though the process itself was uncontroversial and would not result in any major changes to the river.

While the company was grappling with the re-licensing issue, the Environment, Health, and Safety Manager at the plant, Steve Groves, was approached by Dan Sosland, a Senior Attorney at the Conservation Law Foundation (CLF), a New England-based environmental NGO, who suggested using a regulatory negotiation approach. Dan Sosland was highly influential both within and outside IP. He was on the Androscoggin Mill's Environmental Advisory Board, and served as CLF's chief legal representative for the re-licensing of Bowater/Great Northern Paper's two hydropower stations on the Penobscot River. In this case, Bowater/Greater Northern Paper had used a traditional notice-and-comment rule-making process, whereby it tried to gain the licences by persuading the State of Maine to rewrite the re-licensing laws, contrary to the objectives of the EPA.[83] This effort was largely unsuccessful, as the EPA appealed Maine's legislative proposal. Following the successful appeal, the FERC ordered Bowater/Great Northern Paper to make substantial improvements at the two sites. Bowater/Great Northern Paper then tried a collaborative approach with a range of environmental actors, but this was unsuccessful as the differences between the environmental groups and the company were too great. The Conservation Law Foundation felt that a particular section of the river should contain water, for example, while Bowater/ Great Northern Paper wanted the section dry. Although FERC decided to offer Bowater/Great Northern a licence if all the environmental improvements were made, the Conservation Law Foundation appealed the Commission's decision as it felt that further environmental improvements were necessary. This appeal is currently being looked at by the Commission.

The Bowater/Great Northern Paper's case seriously worried regulators, industrialists[84] and even the media.[85] In addition, Sosland and fellow environmentalists (in particular Ken Kimball of the Appalachian Mountain Club) were concerned that this case would set a precedent for similar re-licensing procedures in Maine and therefore sought out Steve Groves (who had a background in state regulation) at Androscoggin to discuss the possibility of using the regulatory negotiation approach. At the same time Sosland (as well as Steve Groves) felt that the state regulators, tired of the long-drawn-out adversarial struggle on the Penobscot River, would also welcome a new type of re-licensing process. Steve Groves was persuaded by Sosland's arguments and, with backing from his managers at International Paper, agreed to try the approach. Groves realized that he was taking a risk, but he felt that he could afford to do so. He had been brought out of early retirement by IP, previously holding a senior post in the State of Maine's Department of Environmental Protection, and had his gamble not paid off he could simply have retired again.

In 1994 he set up a collaborative team with local and national environmental NGOs, representatives from International Paper, the Maine regulatory authorities, and national regulatory authorities (EPA and FERC). The participants were not financially compensated, although they would have a hand in influencing the outcome of the re-licensing process. The collaborative approach was based on a regulatory negotiation similar to that outlined by Harter,[86] comprising the following elements:

- frequent meetings closed to outside observers to allow more open discussion
- confidentiality of proceedings
- a mediator would be involved to guide discussions
- the legitimacy of the process is a form of collective judgement

The team met on numerous occasions over a three-year period (at one stage they met on a weekly basis) and they looked at the following issues:

(a) shortlisting consultants to conduct the scoping studies and prepare the hydropower licence;
(b) identifying background studies (29 were ordered) for the Draft Environmental Assessment as well as the consultants to carry them out;
(c) commenting on the wording of the Draft Environmental Assessment, scoping studies and licence applications;
(d) outlining the concessions International Paper should make on environmental grounds (e.g., lack of a free-flowing stream);
(e) participating in public meetings and site visits associated with the re-licensing procedure.[87]

The collaborative team, with International Paper's backing, hired two consulting firms to handle the re-licensing process. Kleinschmidt Associates, an engineering firm based in Pittsfield, Maine, received the general contract, while a sub-contract was issued to Alec Giffen at Land and Water Associates of Hallowell, Maine. Land and Water Associates provided technical support for the environmental NGOs participating in the re-licensing process (that is to say, Atlantic Salmon Federation, American Rivers, Appalachian Mountain Club, Conservation Law Foundation and Trout Unlimited). In addition smaller sub-contracts were awarded to a large number of consulting firms to conduct the various background studies needed for the Draft Environmental Assessment.

Outcome of the process

As a result of the process, International Paper received its four hydropower licences in record time; the process took four years to complete, and cost around $4–5 million. The company received licences for up to 50 years, 15 years longer than usual, and was allowed to increase the generation capacity at one of its dams, Livermore, by more than 50 per cent. Once submitted, the licence application was commented by the US EPA and the town of Jay which was neutral towards it, as well as by the Conservation Law

Foundation and the Appalachian Mountain Club which came out strongly in favour of the licences being given (these two latter groups bitterly oppose Bowater/Great Northern's re-licensing application). The FERC accepted the application without any requests for further information and, on 21 September 1998, prior to the old licence expiration, re-licenced International Paper's four dams, a first for FERC.

IP's successful result did have a cost. The environmental concessions that emerged from the negotiated rule-making process included:

(a) donating 96 shoreline acres in the Rangeley Plantation to the state of Maine to be added to Rangeley Lake State Park;
(b) negotiating a land stewardship deal, giving 280 acres to the Androscoggin Land Trust and conservation leases of up to 50 years on an additional 956 acres to the Trust;
(c) stocking 250 brown trout annually in the Androscoggin River downstream from the Livermore Dam;
(d) developing or enhancing several recreation facilities along the Androscoggin River (e.g., boat launching facilities).

Environmental NGOs, state and national officials and IP itself were all pleased by the outcome of the process. This was reflected in the media[88] following the licences being granted.[89]

Dan Sosland of the Conservation Law Foundation was quoted as saying: 'International Paper will obtain its licences in record time and at a low cost.' Another commentator on the process said: 'The IP process clearly demonstrates that the federal licensing of Maine's major hydro stations produces significant benefits to the ecology and people of Maine',[90] while the Governor of Maine, Angus King, was quoted as follows: 'An excellent example of how the regulatory system should work...you've got a faster licence, lower litigation costs, land for the people of Maine, and a really spectacular result.' Finally, Steve Groves announced that 'IP saved as much as $1 million in pursuing its licence with a collaborative approach instead of the normal FERC process.'

Following the successful completion of the project, IP's Androscoggin Mill has won both state and national environmental awards. It is now involved with two of the US Environmental Protection Agency's

XL projects (of which there are only 50 in the USA) and it is being considered for a Partnership Excellence Award from the EPA for the success of the re-licensing effort. The paper mill is also the most profitable in the IP group.

The people participating in the collaborative effort saw the following advantages to the approach:[91]

1   It increased public trust in this form of regulatory (consensual building) process.
2   It allowed the team to develop integrated ways of compensating the environment for the 50 year re-licensing agreement.
3   The bulk of the resources involved were directed towards the environment rather than towards lawyers, as would have been the case in a traditional re-licensing procedure.
4   It promoted greater environmental engagement for the licence-seeker. IP was not simply dictated to by state and national regulatory agencies; rather, the collaborative effort, involving a wide array of interest groups, commissioned a series of background studies, fully funded by IP, on topics from the history of exploiting the river to brown trout populations, which were then included in the Draft Environmental Assessment.[92]
5   The process almost completely excluded lawyers who could have increased the lack of trust in the regulatory process as a whole and would have inevitably increased the costs as well.

Analysis: the issue of trust

The negotiated rule-making process appears to have contributed to rebuilding public trust in IP. What were the reasons for this?

First, although there was little public trust of IP, the re-licensing issue was not controversial. The re-licensing of the run-of-river hydropower stations would have little noticeable adverse effect on the environment in the eyes of the local public. It was not a question of siting and building new dams, or a new paper mill. If that had been the case, there would arguably have been considerably greater public opposition.

Second, the main NGO stakeholders involved in the process, Ken Kimball of the Appalachian Mountain Club and Dan Sosland of the Conservation Law Foundation, were not interested in promoting distrust in either the industry or the regulator.[93] Their previous experience in environmental regulatory processes, where distrust had caused problems and less desirable outcomes, had made them open to new approaches. Kimball and Sosland were interested in getting environmental compensation measures in the local region in exchange for a re-licensing of the four dams.

Third, IP was open-minded and willing to work with environmental groups. IP, in this case led by Steve Groves, saw distinct advantages in working with NGOs and regulators as a way of building positive public views toward the Androscoggin Mill, a site known for its environmental and labour controversies.

Fourth, as shown in past case studies as well as in the literature on communication, face-to-face interactions with a wide array of participants can lead to the process being perceived by the participants as fair and trustworthy. Contentious issues can be more easily addressed in an open negotiating setting thereby making the process fairer and, in addition, face-to-face communication makes it easier for participants to assess the trustworthiness of one another.[94]

Analysis of the risk factors

This part of the analysis is grouped around the risk factors highlighted in Chapter 1, focusing on the four main actors in the process:

- International Paper Company, the 'industrialists'
- the environmental NGOs, in particular American Rivers, Appalachian Mountain Club and the Conservation Law Foundation
- the regulators, both State and Federal
- the general public, here represented by the Canton Planning Board and Jay Town Office

In a high public trust, high/low uncertainty risk situation, deliberative risk management strategies are not required. This factor does not apply in this case. IP perceived a high level of public and interest group distrust in the Androscoggin case.

In a low public trust situation, risk management strategies are needed but the choice of strategy depends on the reasons for the mistrust. International Paper was in a difficult position, with a poor environmental and social record. It was mistrusted by both the environmental groups and the local public. To combat the issue of competence, IP first hired Steve Groves, formerly of the Maine Department of Environmental Protection (DEP), to lead on health, safety and environmental issues at the Androscoggin Mill. To address the issue of fairness it set up a deliberative exercise in which federal and state regulators, industry representatives and environmental groups could all participate.

Deliberative techniques can help create public trust in a contentious risk management issue, if the public distrust issue has something to do with partiality, but are expensive and time-consuming. In the Androscoggin case, public trust of IP increased through the use of deliberation in the re-licensing process. This was exemplified in three main ways:

1   The collaborative team held three public meetings regarding the re-licensing procedure. According to observers at these events, not a single individual opposed the process or criticized what IP was doing.
2   The initial hostility of the environmental groups on the collaborative team towards IP and their plans changed to more constructive engagement as the process developed. Following the completion of the process they supported the outcome, saying it was the most favourable for the environment.
3   Several senior IP staff were sceptical about the process. They were particularly concerned that by involving Dan Sosland and other environmentalists, the re-licensing process could run into the same problems as Bowater/Great Northern Paper's attempts to re-licence dams along the Penobscot River. Following the successful outcome they were highly complimentary of Dan Sosland and the model used.

The deliberative process in this case was time-consuming. People were spending at times half a day a week working on the re-licensing project,[95] but it was not expensive. The cost to IP was in the order of $4–5 million for the re-licensing agreement. Perhaps it was no less expensive than a conventional form (adversarial) of re-licensing procedure in which the regulators decide what the outcome will be,[96] yet the process itself (of seeking the licence) was extremely efficient (IP received the licence in record time) and generated significant trust towards IP's Androscoggin operations from the local public as well as state environmentalists and regulators.

In high distrust situations, charismatic individuals are extremely helpful in negotiating successful deliberative outcomes.[97] In many cases these individuals can make or break outcomes. In the interviews conducted for this case it was basically unanimously agreed that without the involvement of both Steve Groves and Dan Sosland the Androscoggin case would not have been as successful as it was. Steve Groves, brought back from early retirement to deal with this and other contentious issues at the Androscoggin Mill, was seen by outside observers as a risk-taker. In this case he pushed for a consensual agreement, even when others within his organization were initially opposed. If Groves had failed in this effort he would have lost his job (and gone back to retirement). On the other hand, if he had adopted an adversarial command and control process and had failed, he probably would have retained his position. As with other colleagues involved in earlier cases, environmentalists would be blamed instead. Dan Sosland, meanwhile, wanted a success story. He wanted to develop a consensual model that could be applied to other hydro re-licensing projects, and in so doing avoid a repeat of the Bowater/Great Northern Paper controversy which reflected negatively on environmental groups.

In any regulatory/risk management process, the political actors, local or national, must be supportive of the final outcome. In the Androscoggin case, all actors, including the political ones, were clearly supportive. Local, state and national actors all rallied round the collaborative effort once the licence was sent to FERC. Both the Conservation Law Foundation and the Appalachian Mountain Club sent in letters in support of IP receiving the licence, and the state and national regulators voiced their support for the licence both in the media and via passing the state permit. In return FERC kept its promise. It delivered the new hydro permit in record time and awarded IP a 50-year licence rather than the standard 35-year one. Hence, all the actors maintained the agreement and public trust in the outcome remained.

It is not enough to assume the regulator or the regulatee has public trust. This must be tested. In the Androscoggin case this was not done. IP assumed it was highly distrusted by both the NGOs and the public, and indeed this view was supported by the actors in the collaborative process. The main reason for this distrust was the prolonged strike that IP suffered and the union-busting activities it implemented. The regulator, in this case the State of Maine's Department of Environment Protection, also realized it was under public scrutiny. As Dana Murch at Maine's Department of Environment Protection argued in an interview on 11 July 2000: 'We had been meeting for too long under adversarial conditions. These conditions led to bad blood and further distrust for all actors involved including the general public. Hence, we wanted to find a win–win, and this case was an example of such a win–win.'

Proactive regulators are more likely to gain public trust than retroactive regulators. With retroactive regulation a policy vacuum could result and be filled by special interest groups. This case study is an example of proactive regulation. The regulators, at both state and national level, were highly proactive and really encouraged IP to adopt this new approach. The outcome of the process was thus one of building trust. The process would not have been successful, however, had representatives from certain environmental NGOs not been asked to participate in the process. In this case they were both enlightened and interested in working with industry to find a win–win agreement, and hence trust in both the process and outcome remained. If an anti-industry environmental group had been asked to participate, the outcome would be significantly different. As Alec Giffen argued (in an interview on 11 July 2000): 'This IP model had the right ingredients. We had enlightened and open-minded NGOs, a risk-taking industrial representative and regulators interested in being proactive.'

Interest groups will in many cases try to create public distrust of regulators which in turn will lead to failures of the risk management process. In most cases interest groups by their very nature want to promote their own interests (be it halting the siting of a waste incinerator or protecting the environment in general). One effective way to do so is to promote distrust in those institutions that are in charge of the issue in question (usually regulators and industry). There are exceptions, however. Appalachian Mountain Club and the Conservation Law Foundation, the two main environmental NGOs in this case, had previously worked with industry regarding re-licensing procedures with win–win results. Arguably, these groups realized that environmental benefits from the collaborative process would be significantly higher through working with industry. According to Alec Giffen:

Had we gone through a standard command-control adversarial process in this case, in which the regulator decides what the company needs to do in order to get an environmental improvement, there is no way in the world that the regulator would have forced IP to give away 280 acres of forest property or set aside 956 acres in a 50-year managed trust. At most IP would have been asked to add some recreation facilities such as boat ramps and picnic tables as well as ensuring a certain flow, but that would have been it.

(Interview with author, 11 July 2000)

Interest groups are needed, however, when the regulator is not seen as impartial and when one is dealing with national or international regulatory issues. In many cases the regulator is not seen as impartial, as in this Androscoggin case. In the Bowater/Great Northern case, the state regulator was highly partial, regarded by environmentalists as working directly with industry in opposition to them. Androscoggin was a local case, neither national or international; the public should have had greater involvement (as in the German case study), but environmental groups were much more involved. Does this imply that the above risk criteria are incorrect?

The difference between this and the German case in Chapter 3 is that re-licensing procedures are complicated and drawn out. With Androscoggin, interest groups occasionally met half a day once a week for nearly 3 years, discussing re-licensing related issues from stream flows to pollution discharges from the paper mill settling in ponds. Most often the general public was unwilling to participate in such processes due to time and finance constraints.[98] In addition, both the state regulators and the representatives from the local towns argued that they effectually represented the public in the process.

5
Sweden: Barsebäck, Risk Management and Trust

Overview

This case study examines the Swedish Nuclear Inspectorate's and Sydkraft's successful communication and management strategy in the aftermath of an INES 2 (nuclear accident of the second lowest severity) incident at the Barsebäck nuclear power plant in southern Sweden in 1992. Sydkraft (the owner of the nuclear plant) and special interest groups, such as Greenpeace Denmark, felt that the Inspectorate handled the situation, that of a near-miss meltdown of a reactor after a filter blockage, properly. The example is all the more striking because a technocratic form of risk management was used, with virtually no involvement of the public or special interest groups in the policy-making process. In addition, a top-down form of risk communication strategy was put in place rather than a dialogue form.

Sweden's overall risk management approach

The overall risk management approach of Sweden is not dissimilar to that of Germany or the UK, in that it is corporate in nature and based on public consensus.[1] As Anton once wrote, the approach is:

> extraordinarily deliberative, involving long periods of time during which more or less constant attention is given to some problem by well trained specialists. It is rationalistic, in that great efforts are made to develop the fullest possible information about any

given issue, including a thorough review of historical experiences as well as the range of alternatives suggested by scholars in and out of Sweden. It is open, in the sense that all interested parties are consulted before a decision is finally made. And it is consensual, in that decisions are seldom made without the agreement of virtually all parties to them.[2]

In Sweden there is very little public deliberation in the policy-making process. Representatives of traditional and established stakeholder interest groups, however, including trade unions, industry representatives and regulators, do participate in policy-making, and it is assumed that these groups act in the interests of the public they represent. Second, there is a strong technocratic tradition of regulation with frequent use of experts to advise on complicated risk management issues, ranging from building a train tunnel through an environmental preserve to analysing the costs of phasing out the country's 12 nuclear power stations. Rational risk policy *does* play a role, although not as strong as in the USA where regulatory impact analyses have been in place since the Nixon Administration. Third, the precautionary principle, whereby regulators argue for a reversed burden of proof, is a central part of the risk management strategy.[3] Fourth, to cope with scientific uncertainty, safety criteria are inherently conservative, resulting in average occupational health standards for Sweden being among the highest in the world.[4]

Sweden's risk management approach can be summarized as shown below:

1   It has a strong technocratic bias with little public participation in the policy-making process.
2   The courts have a limited role in either making or testing regulations.
3   There are close links between the elite stakeholders (called *Överhet* in Swedish) ensuring that consensus emerges, resulting in little conflict in the regulatory process.
4   Levels of public trust in the regulatory system are high as the public sees the regulators as acting in their best interests.
5   There is a strong focus on the precautionary principle (reverse burden of proof) which is agreed through negotiation between the regulated party and the regulator.

6 There is widespread acceptance of the *överhet* by the public, entrusting it to make the right decisions for the country as a whole. The public sees the *överhet* as having made a significant contribution to its welfare.

7 The *överhet*, and in particular policy-makers, are driven by the agenda of the political party which originally put them in power.

8 Any distrustful interest groups (be they environmental groups or pension organizations) are included in the centralized policy-making process and, in some cases, are directly funded by the central government to reduce their impact.[5]

9 Highly proactive regulators lead to little chance of a policy vacuum.

10 The regulator takes the role of overseer while the regulatee is responsible for all the risk management measures carried out.

Historical background

Swedish regulation is consensual rather than adversarial. The Swedish system of regulation emerged from the *överhet*-state, in which dominant values encouraged individuals to defer to the wishes of the government and encouraged leaders to be self-confident in charting a course for how people should behave.[6] This *överhet* has emerged over the last 500 years, and remains strong. One of the characteristics of the system is that policy-makers are seen as individuals who shape public opinion; this is in sharp contrast to the USA where some researchers argue that policy-makers are hostages to public opinion (reactive policy-making).[7] The roots of inclusive consensual decision-making can be traced back to the partial democratization of Sweden in 1865 (20 per cent of the male population was allowed to vote), when the traditional class system, referred to in some circles as the pre-democratic patriarchal state, fell apart. In its place a new *överhet*, composed of local and national policy-makers, senior civil servants, and senior industry spokespeople emerged. These groups formed the so-called new Swedish elite. As the country is small and homogeneous, this rising class tended to interrelate personally, read the same left-of-centre newspaper (*Dagens Nyheter*), frequented the same social circles and was educated in the same universities.[8] Many up-and-coming policy-makers therefore shared a similar background; most joined the government after extensive periods in their respective party's youth society (more than half of the present Swedish cabinet were members of the Social Democratic youth organization, SSU); and they are referred to in colloquial Swedish as 'political broilers' (i.e, chickens raised for politics rather than meat). The political affiliations of the policy-makers are formed very early in their childhood, and are traditionally passed from generation to generation, ensuring that most individuals today vote in the same way as their parents and grandparents. Because of their early involvement in the political youth organizations, Sweden's current policy-makers may have known their counterparts for most of their lives.

Just as the old *överhet* maintained power, so does the *överhet* today. As Tom Anton once said, 'In Swedish politics, to accommodate is to survive.'[9]

Adversarial as opposed to consensus decision-making

There are few confrontations in Sweden. In comparison to federal states such as Germany and the USA, power in Sweden is centralized and based on consensus, resulting in consistency and leading to effective decisions that suffer few challenges once proposed.

Thanks to an inclusive approach to decision-making, Sweden has one of the lowest occurrences of industrial disputes in the western world,[10] as most problems that arise are dealt with centrally by the main players, such as the large trade union bodies (in particular LO), the employers' federation (SAF) and government. For more than 50 years these bodies have negotiated annual wage increases without a strike.

For example, Sweden has the highest percentage of unionization in the western world with 85 per cent of the work force belonging,[11] and as a result union bodies such as LO and TCO have considerable amounts of power. Because of this, Lennart Lundqvist argues, participation in the policy-making process is based on competence and not citizenship.[12] Indeed, the public by and large appears to have accepted this, and on the whole people do not take an active part in trying to influence the policy-making process.[13]

Regulations that are developed centrally, involving SAF and other industry representatives, various union officials, and government agencies, are thus simply imposed on industry. Although at the local level there may not be an agreement regarding a particular regulatory decision, it is highly unlikely that it will be challenged. Historically, there has been a sense that the regulator will work with industry to ensure compliance rather than seek judicial intervention or impose fines, as with the country's regulation of vinyl chloride in the 1970s.[14] In addition, unlike the USA, where regulation is by the book, Swedish regulation is inherently flexible, implementing a dialogue between the regulator and those regulated as to how risks are best managed. In this process, the regulated party is responsible for putting forward risk management alternatives and implementing them, while the regulator is seen strictly as an overseer.[15]

As the regulatory process is more flexible, if not less bureaucratic, changes can be made at a moment's notice if new scientific evidence becomes available. Regarding the regulation of vinyl chloride in the 1970s, for example, Sweden first adopted a 5 parts per million (ppm) limit but, following evidence from the USA stating that limits above 1 ppm could be carcinogenic, the Swedish regulators swiftly adopted similar standards. In this case the only industry plant that produced these chemicals in Sweden was unsure that it could cut back to 1 part per million. During the regulatory process, however, the director of the industry in question was assured by the regulatory counterpart (they were close friends) that the regulator would not shut down the industry and would help it reach the desired limit.[16]

A final explanation of the high degree of consensus in Sweden relates to the relationship between the public and the government. Unlike the USA, where the main interaction with national government is through the tax system, there is a significant fiscal link between people and government, and a bigger government presence in most areas. Sweden, for example, has a big public sector, with half of all Swedes dependent on the state for their salaries.

In the area of environmental and safety regulation, regulators, industry and other principal actors have similar goals to develop regulations centrally with little confrontation by involving all the powerful stakeholders. A good example of building consensus can be found in the Swedish environmental policy arena where the first Director General of the Swedish Environmental Protection Agency, the legendary Valfrid Paulsson, adopted a strong cooperative relationship to form greater consensus between the regulator and the regulated. Paulsson, who was the Director General for 25 years, took the view that a working relationship between the regulator and the regulated had to be nurtured. As Lundqvist argues, the formation of such a climate is only possible via information and trust:

> One key was information; if only each side knew exactly what the other wanted and why, the relations between controllers and polluters would lead to rational and balanced decisions. Another was trust; polluters should be relied upon to execute agreed pollution control programmes and prescribed control measurements, without day-to-day interference from regional or local environmental officers.[17]

The role of trust between regulator and those regulated is also shown in the emphasis on self-regulation. Companies in Sweden conduct their independent health and safety audits so as to improve safety awareness at all levels of the organization. The strength of this approach remains to the present day.

Consensus leads to slow decision-making

The regulatory approach adopted in Sweden is on average rather slow in its formation, as it is incrementalist and long term in nature.[18] To ensure consensus, all the major decision-makers interact, as Thomas Schelling argues, through the creation of several fora for negotiating with interested parties. When decisions are made, every participant feels part of the process and the outcome is psychologically rewarding.[19]

Swedish policy-makers do not want to have what they call 'a cow on the ice' (literally translated, but meaning a large element of uncertainty): that is, they do not want to be driven by public opinion to such an extent that the implementation of environmental or safety regulations may lead to a sacrifice of various socio-economic variables. Doing so could have detrimental effects on the main cornerstones of Swedish policy, namely full employment, stable socio-economic growth and welfare. One way of reducing the element of policy uncertainty is to narrow the political focus via the establishment of Royal Commissions, which consist of administrators and representatives of regulated interests. These commissions, together with the cabinet and bureaucracies, help decide what is feasible practically with regard to a particular regulation. Once a regulatory decision is passed it is always viewed as general, flexible and by no means final. Regulators want to ensure that the legislation passed will not simply focus just on a specific pollution source from a particular plant, but will be more holistic in its orientation so that it can be applied to a variety of pollution problems and cases. It *must* be flexible so that the regulator can implement his or her discretionary powers to assess the balance of costs and benefits involved in the regulation case by case.

Role of the public

Until recently, public participation in the policy-making process was discouraged by local and national policy-makers. Public opinion is taken into account through the so-called remiss process whereby, following the passing of draft legislation, the wider community (be they special interest groups or private citizens) is allowed to pass comment before it becomes law. Policy-makers may act on these comments in the final drafts of their bills, but this usually depends upon the nature of the comments themselves and the perceived influence/power of the body making them.

Policy-makers argue that state regulatory bodies *must* act in the public's best interests, such as by advocating a safer work place and a cleaner environment. Therefore the public does not really need to participate in decision- or policy-making at all unless employed by the government. Indeed, policy-makers view public participation as leading to inefficiency, thereby prolonging the regulatory process. The public appears to accept this situation just as it did the old *överhet* and the powers that be. Indeed some observers acknowledge that over the years a new aristocracy has formed, which is based on political connections rather than blood ties.[20] This should not be surprising, as in terms of outcome the Swedish public are no worse off and may be considerably better off in terms of regulatory decisions than their counterparts elsewhere.[21]

The political process

Sweden's current parliamentary system, based on proportional representation, also encourages strong party loyalties. Several commentators argue that this makes policy-makers less responsive to public pressure on emotive issues.[22] The Swedish political system is generally classified as a stable multiparty system with a high degree of internal party cohesion.[23] The parties usually vote in blocs, although exceptions have occurred in the past. The Conservatives, Liberals and the Centre Party usually vote together and form a centre-right bloc. Similarly the Social Democrats, the Communists and the Greens form the Socialist bloc. The political bloc with the most votes wins, and within that bloc the party with most votes is asked to form a coalition and establish the government. As a result, the MPs are not directly beholden to public opinion and can therefore liaise with the other members of the *överhet*. Because of the lack of direct public accountability, some policy observers argue that on the whole Sweden is considerably less democratic than other European countries and the USA.[24] That said, voter turn-out in Sweden is almost twice as high as in the USA. Although there is built-in flexibility in the policy-making process, there is a sincere wish among Swedish policy-makers to make rational rather than simply populist policies.

The limited role of the courts

The law courts, to date, have played only a limited role in regulatory policy-making. Judicial review of administrative performance is extremely rare as administrations are seen to act in the interests of the country. Although there is a right of appeal regarding regulatory decisions in a court of law, this only applies to formulaic errors. Hence citizens and interest groups cannot use the courts to challenge government policy (something that environmental NGOs in particular oppose).[25] This has changed somewhat since Sweden joined the European Union. Swedish citizens or organizations can now appeal to the European Courts, as the utility company Sydkraft did recently, regarding the government's decision to close one of the company's nuclear power plants.

Proactive regulation

Sweden's regulatory system can be characterized as relatively proactive. The government and regulators actively draw upon new scientific research to develop existing and forthcoming legislation. The precautionary principle and shifting the burden of need/proof to industry are both part of the regulatory process. In some cases, industry has not introduced potentially hazardous chemicals to Sweden, as it has taken the view that the costs of satisfying the regulator (i.e., proving that the chemical substances are indeed safe) are too high in relation to the small market Sweden represents. As a result, the country has fewer chemicals in circulation than most other western nations.[26] Swedish regulators have asked a Criteria Group to reconstruct unsystematic (and even, at times, non-existent) implicit safety margins for chemicals. The outcome of these reconstruction experiments has, in some cases, led to the imposition of conservative safety margins so as to reduce scientific uncertainty about safety. Sweden has the lowest occupational exposure limits to chemicals in the world.[27]

Introduction to the Barsebäck case

There are several important factors that contributed to the broadly satisfactory outcome for all participants and stakeholders, and this chapter explores how a risk management approach which avoided direct involvement with special interest groups actually succeeded.

What happened?

The controversy centred around the Swedish nuclear (boiling reactor) plant located 20 km from both Malmö and Copenhagen, near the small harbour town of Barsebäck (see Figure 5.1), which first came on line in the 1970s. The plant is situated close to densely populated areas (Copenhagen has approximately 1.5 million inhabitants, while Malmö has approximately 350,000 inhabitants) and has therefore been a source of public and political friction between Denmark and Sweden for over 25 years. On 28 July 1992, during the start-up of reactor II at the Barsebäck plant and following routine maintenance, a safety valve became stuck, causing insulation material to fall into the reactor's water-cooling system, blocking the reactor's inlet filters. Within 20 minutes, however, this material was cleared through a back-flooding mechanism. Nevertheless, the incident prompted the SKI (the Swedish Nuclear Inspectorate) on 17 September 1992 to force a shutdown of all similar filters, arguing that any blockage increased the risk of a complete core meltdown. It also ordered the owners of the five affected reactors to redesign their filter systems.

The SKI regulator was particularly concerned about an incident violating the so-called '30 minute rule' which compromises the safe operation of the nuclear power plant. Studies have shown that most mistakes in dealing with nuclear-related accidents or incidents take place within the first 30 minutes when operators are under acute stress. The rule therefore states that an operator of a nuclear power

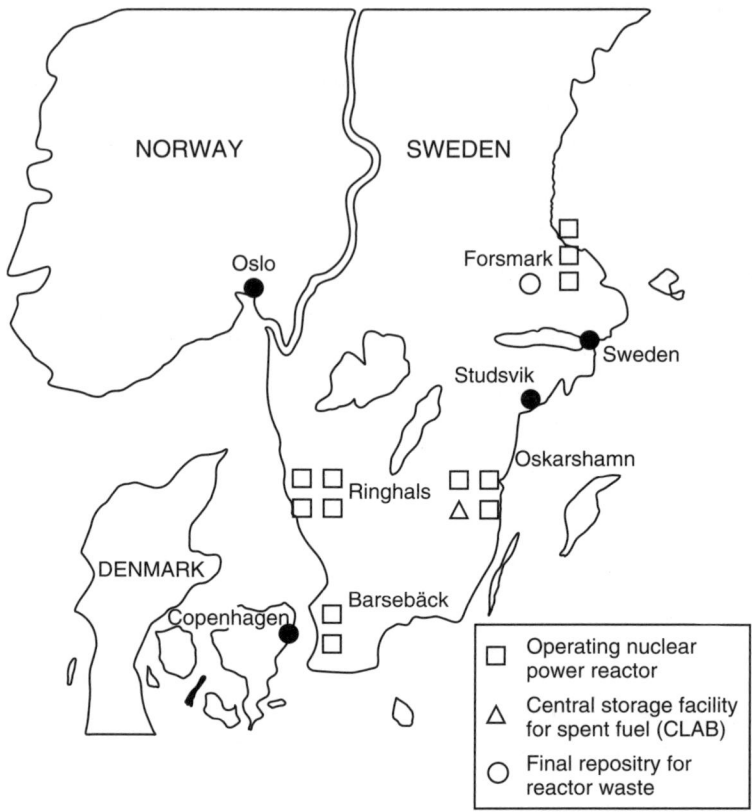

Figure 5.1   Sweden's nuclear reactors

plant should always be allowed 30 minutes to come to terms with the problem at hand, thus ensuring that he or she responds correctly to the incident in question.

The Danes felt that the reactors should not have been constructed in the first place, since the plant not only put Sweden's third largest city, Malmö, at risk, but also the Danish capital with its 1.5 million inhabitants. Danish officials discovered that Copenhagen's evacuation plans were insufficient, estimating the costs for dealing with the accident at nearly 50 billion Danish crowns (approximately 6 billion euro). This prompted the Danish Parliament to pass a bill in May 1986 calling for Sweden to close Barsebäck.[28] Yet pressure for closure fell on deaf ears in Sweden, as the Swedes saw the plant as being completely safe, prompting the Danish government to raise the issue on a yearly basis ever since.[29]

Opposition to the plant could also be found in Sweden where the Swedish Centre Party (headed by the anti-nuclear campaigner, Olof Johansson), and several other influential Swedes (including Kjell-Olof Feldt, the ex-Finance Minister) lobbied alongside NGOs such as Greenpeace Sweden and Friends of the Earth to close the plant permanently.[30] In effect a single stuck safety valve seemingly threatened to wipe out nuclear power as a viable energy source for Sweden.

Despite opposition, Sydkraft completed modifications to the filtering systems and the reactors went back on line in January 1993. This generated a massive outcry in Denmark. The complete lack of success in finding a diplomatic or political solution agreeable to both countries prompted the Danish Interior Minister, Thor Pedersen, to suggest that his country take over the provinces Sweden captured in 1658 (including Scania [Skåne] where the Barsebäck plant is located) by military means. Carl Bildt, the Swedish Prime Minister, felt this statement to be extremely ill advised, as it jeopardized the foundation of wider Scandinavian cooperation.[31]

While diplomatic initiatives stumbled forward, anti-Barsebäck campaigners stepped up their own initiative with a high-profile 'war of humour'. Diplomatic relations were at such an all-time low that Sweden's Minster of Defence, Anders Björk, threatened to 'attack' the Danes with fermented herring. Journalists from the Danish newspaper *Ekstra Bladet* responded by dumping old smelly cheese at the Barsebäck plant.[32]

The price of safety

The consequences of action taken to avoid danger are instructive. Shutting down the reactors and redesigning the filters may have protected the population in the long run, but short-term consequences for the environment included a 10 per cent increase in Sulphur dioxide (SO_2) and nitrous oxide (NOX) emissions in southern Sweden, causing increased acid rain. This was because the shortfall in nuclear-generated electricity was compensated for by an increased usage of fossil fuels in southern Sweden and Denmark. Renewed reliance upon old Danish coal plants meanwhile increased emission of sulphur dioxide. Studies show that on average up to 23 per cent of sulphur dioxide (the main ingredient of acid rain) falling over southern Sweden originates from Denmark.[33] The Barsebäck shutdown also resulted in a loss of Sydkraft revenue to the order of SEK 25 million per week – with substantial profits flowing into Danish power companies now supplying energy to large parts of southern Sweden.

Before proceeding further, some background to the siting and building of the two Barsebäck reactors is necessary: this is given in Figure 5.2.

The seeds of opposition

Opposition to the Barsebäck nuclear plant was a domestic issue in Sweden and did not erupt into international activism until the early 1970s. In 1965, the utility company Sydkraft was not planning to invest in radical, state-of-the art nuclear power. Instead it envisaged a large oil-fired heating and electric plant, and negotiated the purchase of a large tract of land near Barsebäck harbour in the province

---

1965   Sydkraft buys land at Barsebäck.
1968   Sydkraft seeks permission to build a nuclear plant at the site.
1970   Swedish Government grants Sydkraft a planning and operation licence for a nuclear plant.
1975   Barsebäck I starts to produce nuclear-generated electric power.
1976   Sweden's 'nuclear election'; the anti-nuclear Centre Party comes into power, forming a coalition with two pro-nuclear parties.
1977   Barsebäck II starts to produce power; Barsebäck is discussed as an issue in the Danish Parliament.
1980   National Referendum on nuclear power in Sweden.
1982   First Swedish-Danish review on the Barsebäck plant; Danish–Swedish Barsebäck Committee is formed.
1985   Danish–Swedish Barsebäck Committee reports to the Danish Parliament.
1986   Danish Parliament demands closure of the Barsebäck plant; Chernobyl accident leads to anti-nuclear public outcry in Sweden and Denmark.
1988   Swedish Parliament decides to shut down two nuclear reactors (one of them Barsebäck) by 1996.
1991   Swedish Parliament revokes the 1988 decision.
1992   Nuclear accident occurs at Barsebäck, causing public outcry in Denmark.
1993   Two reactors at Barsebäck are reopened, generating more public and political controversy.

Figure 5.2   Barsebäck history, 1965–93

of Skåne for SEK 9.5 million. But heightened concern over the political situation in the Middle East (the Arab–Israeli conflicts) and its effect on oil prices, in addition to encouragement from Asea (now ABB-Asea Brown Boveri), led Sydkraft to consider nuclear power as a serious energy alternative. The location was viewed as particularly favourable for a nuclear plant since it had a very low population density within the required 5 km safety radius at the time.[31]

During the siting process of Barsebäck, the Danish government was kept informed of Sydkraft's plans by their Swedish counterparts. At that stage, although the proximity of the site to nearby urban centres was close, the feeling among the majority of Swedish and Danish politicians was that the plants did not pose a massive threat, leading the Swedish Government to grant Barsebäck its planning and operating licence in February 1970.[32] By 1975 the first reactor, Barsebäck I, was generating power, and by 1977 so was the second reactor, Barsebäck II.

Between the initial planning application in 1968 and the granting of permission to build in 1970, Danish policy-makers raised no objections to the reactors. In fact, in the early 1960s, a group of Swedish utility companies, with the full collaboration of the Swedish government, built a small nuclear reactor virtually in central Stockholm (Ågesta).

Studies of safety *were* being conducted. The Danish government received a report from the Risö laboratory and the Danish Atomic Energy Commission (the Risö safety document) which discussed the possibility of aircraft from nearby Copenhagen airport crashing into the Barsebäck plant.[36]

The lack of Danish opposition towards Barsebäck may also be partly explained by the country's own plans for nuclear power. Until the mid-1970s, Danish utility companies considered building nuclear reactors of their own.

Barsebäck and Swedish opposition

Towards the mid-1970s, Barsebäck began to face strong domestic opposition from the Swedish Centre Party. At that time, nuclear power entered the headlines as members of the Centre Party showed concern about the treatment of the nuclear waste being produced, something that the public became increasingly worried about as well. Although public opposition to nuclear power was also seen as a protest vote to the centralized system in Stockholm,[37] the issue of nuclear power became so politically charged that the country polarized into yes and no camps, culminating in the so-called nuclear elections of 1976. A centre-led coalition gained power for the first time in 44 years, mainly due to its anti-nuclear platform. The Centre Party leader, Torbjörn Fälldin, pledged at his election that all nuclear plants currently in operation would be phased out by 1985, and that those still under construction would never be completed. His pledge was soon abandoned, but after the 1979 Three Mile Island nuclear accident in America, which led to a national referendum on nuclear power in 1980, the Swedish government passed a law calling for a complete phasing-out of nuclear power by 2010.[38]

What is remarkable is that only a year after the referendum, opinion polls showed antinuclear sentiment to have died down. An antinuclear rally at the Barsebäck site in 1982 only attracted 6,000 people, compared to 20,000 a year prior to the referendum (March 1980), and the nuclear power issue was thought to be resolved.[39] With the exception of a brief period six months after the Chernobyl nuclear accident in 1986, when 85 per cent of the public opposed nuclear power, Swedish opposition to nuclear power has remained a marginal rather than a mainstream political issue.

The Barsebäck plant and the Danes

It is evident that Danish political opposition to Barsebäck has grown over time. Safety concerns were voiced by Preben Wilhjelm in the Danish Parliament in October 1977. Worried about the start up of Barsebäck II (in the light of the Risö safety report), he challenged the Environment Minister to explain how the government planned to protect Danish territory against the risks posed by Swedish nuclear reactors. The Minister, ignoring the safety aspects raised in the Risö document, replied that there was no reason to complain to Sweden about the safety of the reactors until the Ministry received additional information. Parliamentary debate on Barsebäck, however, continued. Following increasing pressure from smaller opposition parties, the Danish government in December 1979 announced that they were seeking a joint Swedish–Danish review of Barsebäck. This was completed in early 1982 and suggested that the plant was nominally safe from risk. Nevertheless, the Danish Environment Minister wrote to Sweden's Minister of Labour expressing Danish concern over the plant's proximity to Copenhagen, resulting in the formation of a further committee to carry out a more thorough analysis of *all* aspects of the plant.

The new committee presented its results to the Danish Parliament in March 1985. Among the report's findings, those relating to the unthinkable economic and political consequences stemming from a hypothetical sudden forced evacuation of Copenhagen due to a Swedish nuclear safety alert prompted the Danish Parliament to pass a bill in May 1986 calling for Sweden to close Barsebäck.[40] Since the autumn of 1982, the majority of the Danish public and policy-makers have been opposed to both domestic and foreign use of nuclear power. Studies in 1992 indicate that 83 per cent of Copenhagen's residents and 73 per cent of the Danish population as a whole believe that Sweden should close Barsebäck since it is so close to Denmark.[41] A majority of policy-makers also share this view.

The risk management process that led to the restarting of Barsebäck following the filter case

The risk management process implemented was that typically found in Sweden: a mix of the standard political regulatory regime with considerable doses of technocracy and rational risk policy. The Nuclear Inspectorate was the main regulator, or risk manager, while the other actors – namely Sydkraft, the Swedish and Danish nation-states and Danish environmental groups – tried to influence SKI's management process.

Following the initial incident on 28 July 1992, as per the Swedish model of regulation, Sydkraft's central safety committee at the Barsebäck plant took the lead in analysing the specific causes of the incident for SKI, informing the Nuclear Inspectorate that the evaluative studies conducted would be completed by 31 August. SKI agreed with these measures and the time scale. After the completion of the analysis, and in particular the study conducted by ABB Atom (the constructor of the original reactor) for Sydkraft, showing that the 30-minute rule could be dangerously compromised, Lars Högberg, the Director of the Nuclear Inspectorate, ordered the temporary closure of the five boiler reactors with similar types of outlet valves in Sweden. In an independent evaluation conducted by Göran Steen, chief of the Swedish State's Accident Commission on behalf of SKI, regarding SKI's handling of the 1992 incident, two findings are noteworthy: the Inspectorate was criticized by Steen for allowing Sydkraft to conduct the investigation, stating that this should have been done by the Inspectorate. To the evaluator this would have been more efficient time-wise. In addition, the evaluator expressed concern that immediately after the incident the reactors operated between 28 July and 17 September 1992, despite the fact that they were deemed unsafe. The evaluator did, however, praise the Inspectorate for closing the five nuclear reactors once the analysis was complete.[42]

There are two issues of interest here. First, the Swedish Inspectorate gave its approval when Sydkraft, the company responsible for the incident, volunteered to conduct the in-depth analysis. The Inspectorate *trusted* the regulatee to conduct a proper study of the causes and consequences of the incident, probably believing that the utility company was in a better position than itself to conduct this type of professional inquiry. In addition, the regulator saved considerable amounts of resources (money and human power) which could be used elsewhere by allowing Sydkraft to assume responsibility. Second, once Sydkraft's analysis was complete, the Inspectorate decided, based on the results of the analysis, to close down almost half of Sweden's nuclear capacity for a period of six months. Based on SKI's decision, one can take the view that SKI was right to let Sydkraft do the evaluation, at the same time supervising any action resulting from the evaluation.

Sydkraft and the utilities

It is crucial to understand that in Sweden the drastic measure to close down the five reactors was not publicly questioned by the leading utility companies. This is all the more interesting considering that no other nation implemented such tough measures regarding reactors that were of similar construction and filter design following the Barsebäck incident.[43]

In Sweden, the regulator, the industry and the state all acted in what can be termed 'the best interests of Sweden'. The regulator knew that drastic action needed to be taken; the event was not predicted by the quantitative probabilistic risk analyses (QPRA) put into place by ABB Atom and it violated the so-called 30-minute rule. The reason why SKI's enforcement measures were made possible with little industry opposition was because of its excellent relationships with the various industrial and political bodies based on mutual respect. In particular, a dialogue process put into place between the nuclear utilities and the

Inspectorate, following the US 1979 Three Mile Island accident, helped considerably.[44] In an interview with the Head of Communication at Sydkraft, Stieg-Åke Claesson, for example, he had the following to say:

> Yes, we felt the measures that the SKI put forward were harsh. Yet we knew that in effect the incident was our fault and that had we complained about it, it would have led to further ramifications leading to loss of credibility in eyes of the regulator as well as the public.
> (Claesson, 29 February 2000, interview with author)

Trust and the Swedish regulatory process

The Swedish style regulatory process, in which a majority of the principal actors were trusted, was seen to be fair in this case. Greenpeace, the SKI, Sydkraft and the Swedish public saw the regulators working in their best interests (e.g., taking drastic regulatory action in closing down five nuclear power stations because the regulators were concerned about public safety). Public concern thus abated only six months after the incident. I conducted a random telephone survey of 100 inhabitants of Malmö which showed that 60 people believed the plant to be safe. When asked why they did (open-ended question), 57 replied that they trusted Swedish industry.[45]

Environmental interest groups, such as Greenpeace, also felt that the process was correct since the regulator took such a strong regulatory step against the nuclear utilities. When asked whether or not the public interest was indeed served by this action, a spokesperson at Greenpeace Denmark responded (in September 1993): 'I trust the SKI more than I trust the Danish inspectorate. SKI is truly not in the hands of industry. In face of industry pressure, it took drastic action to close down reactors that it did not feel were safe. That would never have happened in Denmark.'

Sydkraft, although a private company and owned largely by foreign shareholders (in particular the German company, Eon), did what it considered best for Sweden. For example, Stieg-Åke Claesson, Information Director at Sydkraft, said:

> Of course we were not very happy about closing the two reactors. It was a major blow for us. However, we never considered questioning the decision of the regulators. Rather we decided to work

with them to find an optimal solution. In addition we invested quite a bit of resources to build up public trust.

(29 February 2000, interview with author)

The company was also actively engaged in a dialogue process with the Inspectorate prior to the incident and was acutely aware of how high the stakes were. If the company chose to publicly oppose the temporary closure of Barsebäck, this would not only have damaged their credibility in the eyes of the government and other utilities, but might have further eroded public trust in nuclear energy as a whole. Clearly this would not have been in the interest of the utilities: Sweden is one of the few nations in Europe in which a majority of the public are pro-nuclear power, a state of affairs that the nuclear industry does not want to jeopardize. There were 12 nuclear power stations at risk in Sweden and the 1980 referendum had already stipulated their eradication by 2010. The war for and against nuclear power was to be won or lost in the hearts and minds of the public after all.

Analysis of the risk factors

If we examine the case study in light of the risk factors outlined in Chapter 1, we come to the following conclusions.

In Sweden, a high public trust high/low uncertainty risk situation, deliberative risk-management strategies are not required. Barsebäck is an example of this. There was considerable uncertainty concerning the risk in question during the risk-management process (e.g., should the reactors be closed down, and would it be possible to widen the outlet filters, etc.) yet a high element of trust, based on past experience also existed, both towards the regulator and industry. That is why these bodies were already trusted by the general public when these actors went into the (crisis) risk management process following the filter incident. The outcome of the process – the addition of a series of safety measures (widening the filter outlet valves and changing the type of insulation material being used), and then reopening the nuclear reactors – did not change this level of trust in Sweden.

This risk management process, as in most Swedish cases, implemented a strict political regulatory regime, with a high amount of technocracy/expertise and rational risk thinking. Throughout the process there was no direct public participation or interest group involvement in the decision making process.

Had the main actors in the risk management process (that is, Sydkraft and SKI) pursued deliberative strategies, the outcome of the overall risk management process may have led to a higher level of distrust toward the regulators and industry. Past studies[46] have shown that direct public participation would have led the public to question the need for deliberation in the first place. Similarly, had interest groups been asked to participate, particularly those that were antinuclear in their orientation, they would have pushed their own interests (i.e., a permanent closure of the reactors), arguing throughout the process that neither industry nor the regulators could be trusted (see Chapters 1 and 2).

In a low public trust situation, some form of risk management strategy (strategies) will need to be implemented, but this depends fundamentally upon the reasons for distrust in the first place. This does not apply to this case, as there was a high level of public trust.

Deliberative techniques can help create public trust regarding a contentious risk management issue, if the public distrust issue has something to do with partiality; but these techniques are expensive and time-consuming. This does not apply to this case since deliberation was not regarded as necessary by either the regulators or the general public. Decision-making is inclusive in Sweden, minimizing the risk of public involvement, and therefore public distrust.

In high distrust situations, charismatic individuals are extremely helpful in negotiating successful deliberative outcomes. This does not apply to this case, as there was a high level of public trust.

In any regulatory/risk management process, the political actors (local or national) have to support the final outcome. In the Swedish consensual/ elite model, all the actors work for one another. In this case the regulator and industry were, on the whole, assured of local and national support for the decisions taken. Policy-makers were consulted prior to the decision being made public, and their concerns and reservations were taken into account in the making of the legislation in question.[47]

It is not enough to assume the regulator is trusted; the regulator must test this first. In the Barsebäck case, the Swedish regulator did not test to see whether it had public trust, the most obvious reason being that it did not have time, and testing for public trust was probably furthest from SKI's mind when the incident at Barsebäck occurred. It could have tested the public's trust toward SKI prior to the event but, as with other elements of Swedish elitism, it assumed that the public would somehow realize that it was acting in the public's best interest.

Proactive regulation is more likely to gain public trust. As discussed in this chapter, Swedish regulators, largely due to the elite/consensual system, are seen to be proactive and hence were able to avoid so called fire-fighting. Although the regulators and industry reacted retroactively in this case (since an accident had already occurred), they quickly came to grips with the situation, putting forward strong regulatory measures that were accepted by all parties involved. As the decision was taken rather swiftly, with little hesitation or debate, and as there were no opponents to the decision (even those who labelled themselves 'anti-nuclear' viewed the decision to close the five reactors as 'good' as well as tough) there was no need for SKI to 'justify' its decision. As an added element of security it appointed a consultant, Goran Steen, to evaluate SKI's risk management process in this case. Although expressing some reservations, the consultant was on the whole positive.[48]

Interest groups will in many cases try to create public distrust of regulators which in turn will lead to failures of the risk management process. This does not apply to this case. The principal actors, such as trade union and industry representatives, actively participated in the risk management process. The actors who could be viewed as being opposed to nuclear power and who may have wanted to see the nuclear reactors closed permanently, however, such as various environmental activist groups, were, as in previous risk management cases, only peripherally involved.[49]

Interest groups are needed, however, when the regulator is not seen as impartial and when one is dealing with national or international regulatory issues. In this case a majority of both industry and nuclear opponents viewed the regulator as impartial and hence there was no need to involve interest groups.

6
Risk Management in the UK: The Case of Brent Spar

Overview

This case study examines the communication and management strategy of both Shell and the British Department of Trade and Industry (DTI) during the proposed dumping of the redundant oil storage buoy, Brent Spar, in the North Sea in the spring of 1995, and its occupation by Greenpeace demonstrators. A technocratic form of risk management was used, with virtually no involvement of the public or special interest groups in the policy-making process. In addition, a top-down form of risk communication strategy was put in place rather than a dialogue form. It is an example of an unsuccessful technocratic approach. Both in the UK and elsewhere the public sided with Greenpeace against the DTI and Shell. These results, however, are not particularly surprising. Following a series of scandals running from salmonella in eggs to mad cow disease, the British public has little trust in government regulators or of industry as a whole.[1]

Introduction: the regulatory context

The British regulatory environment is seen as consensual, retroactive and incremental: regulators work closely with industry and special interest groups. This approach is well summarized by Hayward, who argues:

> Firstly, there are no explicit, over-riding medium or long term objectives. Secondly, unplanned decision-making is incremental.

Thirdly, humdrum or unplanned decisions are arrived at by a continuous process of mutual adjustment between a plurality of autonomous policy-makers operating in the context of a highly fragmented multiple flow of influence. Not only is plenty of scope offered to interest group spokesmen to shape the outcome by participation in the advisory process. The aim is to secure through bargaining at least passive acceptance of the decision by the interests affected.[2]

Or, as Macrory argues: 'Discretion and practicability might be described as the hallmarks of British environmental law and policy, with a degree of satisfied isolationism and administrative complacency running close behind.'[3]

However, recalling the ideal types outlined in Chapter 2, the British regulatory approach has, like the other countries surveyed in this book, a strict regulatory regime with components of litigation attached to it. Like Sweden, it also has a strong technocratic (expert component) dimension, consisting of expert scientists giving enlightened civil servants advice as to how to best manage risks. Rational risk management on strict economic grounds does play a role in the UK, but the use of cost–benefit analysis and other such economic tools is not as pronounced as in the USA.[4] Until recently, the British regulatory regime had very little room for either public or interest group deliberation, yet because of growing public distrust in the regulatory process this is changing.[5]

Historical context[6] (1840s–circa 1990)

The consensual approach in Great Britain dates from the peak of the industrial revolution in the 1840s–1860s, when industry abandoned its controversial, aggressive stance which typified the early part of industrialization. This reflected a change in the political climate. In 1842 the government had already introduced the 'best practice' principle as a way of decreasing its involvement in industry and letting industry regulate itself, which it welcomed. (The more modern concept 'safe as reasonable practicable' was first coined in the 1949 court case of *Edwards* v. *The National Coal Board* when Lord Asquith, presiding over the case, referred to it.)[7] Similarly in the early 1860s, Parliament, concerned about increasing levels of air pollution, pushed for stricter regulations which the mill owners initially opposed. In 1863 Parliament passed the Alkali Act which required the manufacturers of akali to remove 95 per cent of the hydrochloric acid produced by their factories. To enforce this act the Parliament established the government's first pollution control agency, namely the Alkali Inspectorate, the world's first environmental regulatory agency.[8] The industry owners' initial reaction to this Act and the setting-up of the Inspectorate was hostile. Working with industry, however, the Inspectorate was able to reduce 98 per cent of the hydrochloric acid emissions and, in so doing, it showed industry that regulation actually reduced their costs.[9] This was, of course, welcomed by industry, which then worked on establishing close links with the Inspectorate. For example, as late as 1981 the then Chair of the Chemical Industries Association's Environment Committee noted the following:

> In the UK we are very fortunate in having relatively easy and frequent access to Civil Servants in both the Scientific and Administrative Branches and many opportunities, both formal and informal, to state a case and to influence the opinions of policy-makers. By and large, our inspectors are practical, professional people, generally able, willing to take account of economic and employment factors so long as willingness to progress is being shown and progress is made.[10]

In addition, in 1874, Parliament passed the Factory Acts, which allowed local authorities to regulate industry themselves. Some scholars, most notably Martin Weiner, suggest that there was a cultural backlash against the values of industrialization. Instead of 'free-market' competition and expansion, and the reactionary opposition to any form of government regulation (indeed this 'entrepreneurial', often 'middle-class', ethos was responsible for much of Victorian England's wealth and power), industrialists, coopted by a resurgent aristocracy, sought social acceptability and economic stability over increasing profits. In addition, as the industrialists grew in wealth they too acquired large country estates, and through this process saw for themselves the destruction of the countryside by industry.[11]

Some of the industries recognized for themselves the long-term benefits of these regulations, and in other cases inspectors were able to persuade manufacturers that this was indeed in their economic self-interest (as occurred following the set up of the Alkali Inspectorate). In addition, because of the change of view within industry, regulators realized that there was less need to police industry, and that it was possible to work with it to find acceptable solutions for all actors involved.

This was made easier by the perceived prestige of the British civil service, and a strong, if not socially overriding, tradition of political authority. At this crucial stage of industrialization, when industrialists developed a conscience and wanted to become more socially acceptable, they found counterparts among civil servants who helped to guide them. The public at large accepted this arrangement as they now viewed the industrialists as 'gentlemen desirous of doing what is right'.[12] For much of the twentieth century, this consensual style of relationship grew stronger. Throughout the 1980s, for example, senior staff of the National Farmers' Union met their counterparts at MAFF almost daily to discuss policy.[13]

Development of current practice

This collaborative arrangement, with civil servants in the various inspectorates becoming in effect free consultants for industry on how to best implement pollution control measures, lasted until the early 1990s.

The consensual regulatory approach worked when there was trust in the system: that is, when the public believed industry leaders could and would work for interests other than their own. This was the case until the late 1980s when there was an upsurge of regulatory scandals. On the other hand, many industrial leaders considered civil servants highly qualified to help them with their specific pollution problems without the risk of being persecuted, and hence promoted the regulatory system as much as possible. As one study on British Water's pollution control policy noted:

> a policy of strict enforcement would destroy the spirit of co-operation with dischargers, painstakingly nurtured by Authorities for decades... although increased enforcement through the courts may yield short-term benefits of reduced discharges, in the longer term it could destroy firm–Authority cooperation, resulting in a decrease

in information flow and an increase in the aggregate costs of reaching a specified water level.[14]

Public trust in the regulatory system lasted for a long time. As late as 1979, only 4 per cent of British respondents in a survey on public perceptions toward regulators felt that a close collaboration between industry and the regulator was improper.[15]

As a result of this 'regulatory relationship' there was little cause for litigation. In fact, inspectors see litigation as a form of failure, believing persuasion to be better in a confrontational situation. It is also difficult for third parties to put forward litigation measures regarding particular regulatory policies, since regulations in the UK are administrative decisions which are, at least in theory, accountable to Parliament and not the courts.[16]

Unlike some forms of technocratic regulatory approaches, the consensual approach is not completely elitist. Both industry and civil servants feel that the best way to achieve a low confrontational environment is to have informal relationships between the regulator, the regulatee and certain so-called interest groups. These last are sometimes referred to as 'legitimate interests', which include several of the leading NGOs and other special interest groups. They can participate as long as they 'behave', acting in such a way as to achieve the best outcome rather than furthering their respective organizations (such as by leaking confidential minutes to the press). As a result, organizations such as the Royal Society for the Protection of Birds, the Royal Society for Nature Conservation and the National Trust have almost received the official stamp of approval in the UK.

This regulatory process was, by most accounts, highly satisfactory. Britain and America achieved similar standards of safety and environmental improvements, for example (the latter with significantly higher costs owing to its adversarial nature), and the regulators were viewed as competent and trustworthy.[17]

Secrecy lies behind the success of the consensual approach. Under the UK's Official Secrets Act there was no obligation to disclose discussions between the regulator and the regulatee concerning the levels of acceptable risk, so that frank discussions could be encouraged. In addition, and in particular until the mid-1990s, the regulatory agency in question was often reluctant to strongly enforce regulation as this would go against the Conservative government's de-regulation policy. This cosy atmosphere was made possible by the fact that the majority of the inspectors had previously worked in industry before joining the inspectorates. Indeed, in some inspectorates this was a requirement for the job.[18]

The UK consensual model also has some negative features, including an overreliance on self-policing. It works when industrial elitists truly act with some form of 'social consideration', if not conscience, but past research indicates that some of the regulators have occasionally flouted the process. In the British offshore oil industry, for example, in their quest for consensus, the regulators sacrificed safety for industry cooperation, and it has been argued that this led to needless loss of life. Carson, for example, states that the regulatory agencies' attempts to achieve cooperation on safety with oil companies when offshore oil exploration began in the 1960s took the place of a statutory regulatory framework for the industry, and had this been in place at an earlier stage it would have led to higher safety standards. Because of the regulator's quest for consensus, however, a regulatory framework was not in place until the end of the late 1960s. Additionally, the fines assigned by the regulator to the offshore oil industry for non-compliance were minimal: an average of £214.[19] Similar standards, according to some observers, were also rather lenient with regard to regulations on hazardous waste in the 1980s.[20]

The changing nature of British regulation

Over the past fifteen years or so, this regulatory environment has undergone profound changes. There are two major developments and a number of smaller ones. The two major ones are associated with changing levels of public trust and the influence of European regulation on the UK.

The decline of public trust in regulators and the effect of European regulations

Over the past years, national polls point to an increasing decline of public trust in both local and national policy-makers. Levels of trust in the UK have been reduced from 39 per cent in 1974 to 22 per cent in 1996.[21] At the same time as the levels of public trust towards regulators are declining they are increasing towards other groups, most notably environmental NGOs. One MORI survey from 1999, for example, asked the following question:[22]

Thinking now about pollution, which two or three, if any, of these sources would you trust most to advise you on the risks posed by pollution?

Pressure groups (e.g., Greenpeace or Friends of the Earth)	61 per cent
Independent scientists (e.g., university professors)	60 per cent
Television	25 per cent
Government scientists	23 per cent
Friends or family	15 per cent
Newspapers	14 per cent
Government ministers	6 per cent
Politicians generally	4 per cent
Civil servants	3 per cent

In sum, environmental NGOs were considerably more trusted than any one else, most notably civil servants, who were only trusted by 3 per cent of the populace.

Why have these levels of public trust in regulators declined so much in the UK? Research has identified a number of factors, ranging from the destruction of social capital (Putnam), to the role of centralization, to increased public knowledge. The largest cause of the decline in public trust towards policy-makers, however, has to do with the large number of regulatory scandals that the UK has been plagued with over the past 20 years, ranging from the mad cow crisis in which MAFF was seen to have lied to the public, to foot and mouth, to salmonella in eggs. The public simply did not view the regulators as being either competent or fair.[23]

A second important factor has to do with the increased importance of European regulation. Increasingly environmental regulations are actually not developed in the UK but in Brussels. There are many examples of this. Because of the 1987 European Large Combustion Plant Directive, the UK government was forced to reduce acid rain emissions, an issue that the Thatcher government had been fighting against for many years. Similarly, the calls for greater use of environmental impact assessment has also been driven by EU laws. Indeed, senior policy-makers in both the Environment Agency and the Health and Safety Executive have noted on a number of occasions that up to 80 per cent of all UK environmental and public health regulations are developed by the European Commission.[24]

There have been the following changes:

1   The government has become more centralized. The Thatcher government sought to reduce the remit of local authorities by shifting regulatory powers to central government.

2   The Government has reconsidered the role of the public. In a climate of high public mistrust of government and industry the Royal Commission on Environmental Pollution (RCEP) recommended that public values should be more included in the policy-making process.[25] This view is shared by the House of Lords[26] and the Lord Chief Justice.[27]

3   The government has been forced to create a more explicit and transparent framework, in line with EU open government guidelines.[28]

4   The government started to change its legal nature from British Common Law to Continental European Roman (civil law), with its more formal legal system.[29]

5   The government has responded to the neo-liberal agenda. Of particular importance is the push for an internal European market, the growth of neo-liberal tendencies within this market (particularly deregulatory initiatives and the opening-up of previously closed public/private markets), and the overall strengthening of globalization.[30]

6   The government has examined the separation of risk assessment from risk management. Similar to the exercise undertaken in the USA in the 1980s, British regulatory bodies are considering the implementation of a more formal separation of the scientific process (risk assessment) from policy-making (risk management).[31]

7   The government has come under greater scrutiny from an increasingly environmentally aware, highly critical and educated public. Combined with this the regulatory environment is forced to recognize the influence of environmental NGOs, especially at the international level.

As Sheila Jasanoff summarizes:

> British society has changed in profound ways that call for new forms of engagement between citizens and their government... institutions which may have been robust enough in their time will have to reconsider some of their fundamental assumptions in order to catch up with the altered state of things.[32]

Yet at least until early 1995 there was an inherent belief that the unique consensual relationship between government officials, certain selected stakeholders and industrial representatives still existed in the UK. If regulatory bodies, industry officials and other principal actors came to an agreement it was presumed it would not face significant opposition from special interest groups and/or the public. By this time, however, surveys indicated that the public trusted environmental groups considerably more than industry, government or regulators. As one contemporary stated, 'Brent Spar was predictable.'

Brent Spar

In early 1994 two oil companies, Shell and Exxon, needed to dispose of the oil storage buoy, Brent Spar, in the North Sea. The buoy was 141 metres tall, had a capacity of 300,000 barrels of oil and weighed 14,500 tons when empty, of which 6,700 tons were steel and 6,800 tons were ballast (mainly concrete).

The buoy, originally commissioned in 1976, had been unused for five years and was thus seen as redundant since pipelines had been laid on the sea bed from the Brent field to Sullom Voe in the Shetland Islands. The buoy was located in deep water (that is to say, more than 75 metres deep). According to the International Maritime Organization's existing guidelines, disposal of the structure in the ocean was an acceptable option. Shell commissioned 30 separate studies to consider the technical, safety, and environmental implications of its disposal and came up with six different options:

- disposal on land
- sinking the buoy at its current location
- decomposition of the buoy on the spot
- deep sea dumping (but within British territorial waters)

- refurbishment and reuse
- continued maintenance

After a thorough examination of these options, Shell decided to implement the fourth option, mainly due to its fairly low cost with little environmental impact (Best Practicable Environmental Option, or BPEO). The next best option, horizontal dismantlement on land, was nearly four times as expensive (£46 million rather than £12 million) and a high risk for workers (six times higher), with a low but measurable risk of pollution of inshore water in case of an accidental break-up during transport. The remaining options were regarded as either unfeasible or environmentally harmful.

Once the deep-sea dumping option was chosen, Shell commissioned a site specific survey to find a location that met two specific criteria:

- the dumped buoy should not interfere with fisheries or navigation
- the area chosen should not be considered a place for either present or probable future legitimate use[33]

Of the three North Atlantic sites shortlisted, the North Feni Ridge, with a water depth of some 2,300 metres, at the northern end of the Rockall Trough some 150 nautical miles to the west of the Hebrides, was selected.[34]

On the basis of these studies, Shell asked the DTI for permission to dump the buoy in the deep sea in the autumn of 1994; to them, the BPEO. In December 1994, the DTI gave its approval. At this point the dumping of the Brent Spar was not an issue. The Energy Minister, Tim Eggar, said retrospectively: 'It had been very difficult to have a public debate about decommissioning because there was very little interest in the issue', while Shell noted that: 'Decisions were neither made suddenly nor in secret. Material was available to the public and the media but no one showed any interest; granting the licence to sink the Spar was covered in the press but only in a few lines because the media decided the story was boring.'[35]

Under the guidelines of the New Convention on the Marine environment (the Oslo–Paris Convention), the British government notified other European nations on 16 February 1995 of Shell's plan to sink the platform. As no country responded within the 60-day deadline for objections imposed by the Convention (by 16 April), the government issued a disposal licence to shell in the first week of May. On 30 April, however, just before the licence was issued, Greenpeace dramatically occupied the Brent Spar buoy itself, while publicly releasing an inventory of what was on board under the title of 'toxic inventory'. The three page document said among other things that it contained 100 tons of toxic sludge and 30 tons of radioactive scale.[36]

A crisis began to unfold. Brent Spar hit the media with pictures, supplied by Greenpeace, of their activists braving water cannons from Shell's tug boats.[37] On 9 May, the German Environmental and Agricultural Ministries lodged a complaint with the British government that land disposal had not been sufficiently investigated by Shell in their opinion. Since this came after the 60-day deadline British officials disregarded this. Meanwhile, throughout May, Brent Spar remained high on the media agenda thanks to concerted lobbying by Greenpeace and certain political groups (particularly in Germany), including a Greenpeace petition signed by politicians against deep sea sinking. These efforts culminated on 26 May when Conservative political groups in Germany joined Green action groups in asking for a consumer boycott of Shell petrol stations. The boycott was effective in Germany, Holland, and parts of Scandinavia. On 23 May, after several attempts, Shell finally removed the Greenpeace activists from the platform, but this did little to silence the actual public controversy, so that by early June a Greenpeace-funded poll in Germany suggested 74 per cent of the population would boycott Shell.

On 5 June the North Sea Protection Conference took place in Esbjerg, Denmark. It was attended by Environment Ministers from the countries surrounding the North Sea and the EU Environmental Commissioner, Ritt Bjerregaard. At the opening of the conference, virtually all the official delegates (except those from the UK and Norway) condemned the sinking of the platform and strongly criticized the British Environmental Minister, John Gummer. On 6 June, the German Environmental Minister, Angela Merkel, demanded a universal halt of deep ocean disposal, including oil platforms. At the G7 summit in Canada, Helmut Kohl, the German Chancellor, informed the British Prime Minister, John Major, that this was 'not the looniness of a few Greens but a Europe-wide, worldwide trend for the protection of our seas'.[38]

In mid-June the platform was again occupied by Greenpeace activists who on 16 June claimed there were 5,000 tons of oil in the tanks which Shell had omitted to take into consideration. A few days later they suggested that the entire structure, some 14,500 tons, was also 'toxic material'.

Throughout the crisis Shell UK received little public support. Indeed, Greenpeace argued that the public opposed the dumping of the Brent Spar.[39] The British government was active in trying to persuade its European allies that the deep sea sinking of the Brent Spar was in fact the BPEO, but these arguments were largely ignored, as the arguments made by Greenpeace were seen by the allies as more credible. Additionally, Shell UK's position was increasingly untenable due to pressure from Shell Germany and the Netherlands where the company was suffering from very negative publicity. At Shell's 1,728 filling stations in Germany, sales were 20 per cent below average; 200 stations were threatened with attacks; 50 were vandalized; two were actually fire-bombed while a third received some gunfire. In addition, as a result of the Greenpeace campaign, Germans were writing letters to the UK DTI, enclosing money to help to pay for onshore disposal. Some even sent pictures of their children urging the chairman, Dr Chris Fay, to stop the planned sinking for the benefit of future generations. Shell-Germany received over 11,000 letters complaining about the disposal in the span of six months.

Faced with the sheer scale of this opposition, Shell called off the sinking of the Brent Spar only hours before it was due to be sunk on 20 June. It cited economic problems due to the boycott. The decision was taken by Shell International in a meeting at their headquarters in the Netherlands (Hague). It put out the following statement: 'The European companies of the Royal Dutch/Shell Group find themselves in an untenable position and find it not possible to continue without wider support from the governments participating in the Oslo/Paris convention.'[40]

It did not consult the British government until after the decision was made, representing a significant departure from the traditional, consensual style of regulation. Shell in fact embarrassed the government, since the night before this decision John Major had defended Shell's proposed dumping in the House of Commons: 'I understand that many people seem deeply upset about the decision to dispose of Brent Spar in deep water. I believe that is the right way to dispose of it ... the proposition that it could have been taken inshore to be disposed of is incredible.'[41]

The UK government's reaction to the Shell decision was immediate. Michael Heseltine, interviewed on Channel 4's news that very evening, noted: 'The embarrassment is for Shell. They caved in under pressure and the Prime Minister has behaved in an exemplary way. He deserved better from a major British company . . . I don't believe that you give in to these pressures. You merely encourage worse pressures to develop.'[42] Feeling betrayed as well as embarrassed, Tim Eggar stated that Shell should have gone through with the deep sea dumping as it was the BPEO. Greenpeace meanwhile issued a statement applauding the action and announced that it would help Shell to find an acceptable environmental solution.

There were other consequences. The British government felt unfairly treated by their European colleagues, a view shared by some of the British press. Following the incident the BBC distanced itself from Greenpeace, even broadcasting arguably anti-environmental documentaries.[43]

On 27 June 1995, Shell started a damage limitation exercise, a media counteroffensive, aimed at German and Danish consumers. In Germany they took out one-page advertisements in 100 national and local newspapers with the title 'We will change'. These admitted to mistakes and ill-advised policies on Brent Spar, but maintained that dumping at sea was correct on technical and environmental grounds. In Denmark, Shell sent letters to 250,000 credit card holders explaining their policies. In July 1995, Shell asked the Norwegian auditing and consulting company, Det Norske Veritas (DNV), to investigate the accusations made by Greenpeace about the contents of Brent Spar's supposedly empty storage tanks (particularly the claim that they still contained 5,000 tons of crude oil). This independent inventory of Brent Spar's contents was published in the autumn of 1995, broadly confirming the figures provided by Shell. A few weeks prior to the report of these findings Greenpeace admitted it had made a mistake about the quantity of the remaining pollutants, but maintained that the sinking of Brent Spar would be wrong. An independent analysis by NERC's Scientific Group on Decommissioning Offshore Structures, set up to consider the scientific and environmental aspects of the deep-sea disposal of Brent Spar, has since disputed both Shell's and DNV's numbers. It suggests neither these samples, nor the two samples taken since then (four in total as of 1998), are adequate measures of what the Spar actually contained, and expresses surprise that no attempt by Shell was made to analyse Spar's contents accurately prior to taking the decision to sink the buoy.[44]

The risks of deep ocean disposal

According to Shell's commissioned studies, the risks posed by the sinking of Brent Spar were quantified: occupational risk was highest with land dismantling and lowest with on the spot sinking. Environmental risks were also low for deep sea disposal. According to these studies, sinking Brent Spar in the deep sea did not pose any significant environmental problems, although other studies by scientists at the Scottish Association for Marine Science (SAMS) felt that some of the biological studies by Shell's scientific consultants (from Aberdeen University Research and Industrial Services) were far-fetched.

Shell's data showed that the amounts of hazardous materials within the buoy were minimal: estimated at 20 tons of oil within the overflow pipes, 48,000 tons of oil contaminated water (40 mg oil/litre), slightly radioactive scale, some oil remnants and other chemicals. This was less than 1 per cent of the amounts discharged by boats in the North Sea in the course of one year. The NERC expert group concluded that the environmental impact of dumping Brent Spar in the deep sea was the same as sinking an empty oil tanker.[45]

There was also a fear of local environmental contamination in the deep sea where Brent Spar would have been dumped, which was not thoroughly researched. This issue was raised by Greenpeace on numerous occasions which noted, based on data from John Gordon of SAMS and John Lamshead of the Natural History Museum, that the marine life around the North Feni Ridge was particularly rich. Research, post-Brent Spar, did conclude that North Feni Ridge was considerably richer in marine life than Shell originally believed. The area was seen as atypical, as the slopes that plunge from the edge of continental shelves, known as bathyal regions, to the deep ocean, exactly the type of place where Brent Spar was supposed to be sunk, contain a greater range of nematode worms than anywhere else on land and sea.[46] The expert group concluded that with regard to the proposed dumping site, Shell and its contractors were guilty of the following:

- paying too little attention to benthic storms and turbidity currents in the area

- largely ignoring the possibility that the contaminants of the Brent Spar could disperse via these phenomena to other locations
- consistently underestimating the biological diversity of the deep ocean
- not acknowledging the existence of specialist communities that derive energy from chemical anomalies in the deep ocean

Analysis

Brent Spar represents a typically British example of the consensual approach to risk management in which civil servants and industry looked for so-called win–win solutions, but with a result that was disastrous. Derek Osborn, who served as the Director General of Environment Protection from 1990 to 1995, succinctly summarized the episode when he wrote:

> The [Brent Spar] episode has reinforced establishment prejudices about the dangers of allowing policy to be determined by environmental 'emotionalism' and activists. At the same time it has reinforced the belief among the environmentally concerned public at home and in Europe that the UK Government and industry cannot be trusted on environmental matters, that all its negotiating positions must be distrusted, and that co-operation with it in common programmes directed towards sustainable development must be looked at very warily rather than embraced with enthusiasm. In short a very bad lose–lose outcome for all parties.[47]

There are several reasons why Brent Spar led to a lose–lose outcome for all involved, and these will be explored below.

Shell and the British Government did not truly appraise the implications of the action

Both actors were very heavily influenced by cost considerations. Dumping the Spar in the sea would have cost in the order of £12 million, as opposed to £46 million for recycling the buoy on land. Shell could have offset a significant portion of the costs against increased taxes (up to 70 per cent) with up to 60 per cent of the burden falling on the British tax payer.[48] The Government, particularly the DTI where Tim Eggar supported the British oil industry,[49] was also in favour of the cheapest option. It represented a significant saving for the nation[50] yet, by promoting this cost-effective rationale, Shell and the government chose to ignore underlying issues which led to the technocratic option being undermined.

Unfairness

The proposed dumping of Brent Spar in the North Sea was biased towards the UK at the expense of bordering European countries (with the exception of Norway). The British government, in collaboration with Shell, wanted to use Brent Spar as a test case for a loophole in the Oslo–Paris Convention (OSPAR), the key inter-governmental authority regulating marine pollution in the Northeast Atlantic (Arctic to Gibraltar). The OSPAR Convention forbade the general dumping of wastes in the sea except for redundant offshore installations and pipelines. Even these installations could not be dumped if materials on or in the installations could lead to 'hazards to human health, harm of living resources and marine ecosystems, damage to amenities or interference with other legitimate uses of the sea' (OSPAR annex III). Although Shell and the government said publicly that Brent Spar was a unique one-off installation, Greenpeace and many other stakeholders saw the case as a precedent for the dumping of other offshore oil platforms in the North Sea.[51]

Thus, by adopting a pro-dumping stance, the government and Shell were in danger of raising opposition in European member states to the disposal of Brent Spar on fairness grounds. Why would they approve the dumping of large offshore oil platforms weighing 14,500 tons which are owned by highly profitable multinational oil companies, when these same European governments, being curtailed by OSPAR Commission guidelines, were not allowed to dump smaller objects in the sea, such as fishing boats, submarines or industrial waste? This message of fairness was indeed eventually raised by various European governments, awoken by Greenpeace occupying the platform.

Greenpeace's interest in the area

Shell and the British government failed on two accounts in their advanced preparation for disposing of Brent Spar. They neglected to assess correctly Greenpeace's knowledge of offshore dumping of industrial waste, and also vastly underestimated the group's potential influence as an international environmental organization. Greenpeace had been campaigning against dumping radioactive waste at sea since 1978. It also clearly identified the 4th Ministerial Conference for the Protection of the North Sea in Esbjerg (Denmark) and the meeting of the OSPAR Commission both held in June 1995 as venues for political lobbying.[52] Knowing this, the British government and Shell should probably have reconsidered trying out their international test case just prior to these meetings.

Low trust

Probably the most important reason why the consensual approach did not work was lack of trust. The public in the UK and elsewhere trusted neither Shell nor the British government, but did trust Greenpeace. During this period the British government was suffering from extremely low levels of trust. The Conservatives had been in power for almost a decade and were plagued with scandals, ranging from perjury, to money for arms, to the BSE scare, Gulf War Syndrome and general sleaze. The public felt alienated by the government, which it saw as favouring industry interests. Government departments were thus acutely aware that they needed public support and trust if they were to manage risks properly. This was clearly expressed by Lord de Ramsey, the former chairman of the Environment Agency:

> There is one vital ingredient without which we will be unable to operate and that is public support. How do we gain trust of the public so that they view the Environment Agency as a 'good thing'? ... Conflicting information has left people distrusting experts, scientists and, most of all, politicians. Now Brent Spar has taught them that they cannot trust the green groups either, something most of us realized a long time ago – I call them the Intensive Scare Unit.[53]

How, then, could Shell and the government ever possibly develop a consensual risk management plan? When Greenpeace admitted it made a mistake on the amount of oil in the buoy, it suffered some credibility problems in the media. In an editorial following Greenpeace's admission of error, *The Times* argued:

> The environmental organisation's admission that it made errors in estimating how much oil was left in Shell's Brent Spar installation

puts the seal on what was already a disreputable campaign. It is also the latest in a series of scientific errors or misrepresentations perpetrated by the campaign group. If the media, politicians and the public now treat Greenpeace's claims with a lot more scepticism, so much the better.[54]

Unlike the Government or industry, however, Greenpeace did not experience a crisis in public trust. Indeed, Greenpeace maintained that there was an increase in membership following the admittance of fault which more than offset cancelled memberships as a result of credibility problems.[55]

Industrialists still believe that Greenpeace is mistrusted,[56] though evidence does not support this. Opinion polls show the public remained against dumping the platform (57 per cent against, to 32 per cent in favour,[57] compared to a poll before the admission where 71 per cent of the respondents opposed sea dumping while 17 per cent were in favour).[58]

Lack of scientific consensus

The British consensual approach has firm roots in science.[59] It is one of the strongest pillars of the consensual model, as scientists work with industry officials and regulators in making regulation.[60] Science is seen as being above policy. In the Brent Spar case, however, there was no scientific consensus regarding the environmental merits of dumping the Spar in the sea and hence the consensual model of regulation did not work.

The problem for Shell and the government was that they needed a strong scientific case to justify the proposed dumping of the Spar. This was necessary to show that the proposed option was indeed the BPEO and that there would be no adverse affects to the environment, as per the OSPAR Commission guidelines. Since there was no clear scientific consensus, this line of argument was bound to lead to controversy. Scientists with knowledge of deep sea environments, and with different views from those hired by Shell, added their voice to the debate, feeding Greenpeace's media juggernaut. Moreover, the outcome of groups of scientists arguing against another is loss of public trust in the scientific process itself. Interestingly enough, Greenpeace later denied it was part of this scientific pluralism. In recent evidence to the House of Lords, it argued that Shell and the government used science to justify their case, but this did not work because the 'public' saw through it: '"Science" was used to justify the dumping of Brent Spar, but the mis-match between the Government framework and the (more sophisticated and accurate) public understanding of the situation has led to a decreased trust in the reliability of scientific advice.'[61]

The scientific debate was not effectively resolved until the Scientific Group on Decommissioning Off-Shore Structures was set up by the National Environmental Research Council, which reviewed and analysed the pros and cons of dumping the buoy in the North Sea. It was perhaps the publication of the Commission's first report that redeemed the reputation of the scientific community involved in the dispute (chiefly oceanographers, marine biologists and geologists).

In a confrontational atmosphere, with scientists pitted against one another, the consensual style of risk management cannot work. Scientists tend to pride themselves on representing a pure discipline where its representatives do not squabble in the press. In this case, however, the principal actors (the British Government, Shell and Greenpeace) used scientists to justify their different positions. In the USA scientists are frequency called in to argue for one side or the other in a dispute, but this phenomenon was relatively new for the UK, especially since the British public was used to the consensual style of regulation in which any conflict is usually resolved 'behind closed doors'. Following the controversy, Greenpeace was even accused by policy-makers and the media of bringing science, the 'pure discipline', into disrepute.[62]

National versus international

The risk management approach as envisaged by the Alkali Inspectorate and the British chemical industry, two national actors in the 1860s, was based on a consensual relationship where there was mutual respect. The relationship was not set up with the modern multi-national industry in mind where regulatory bodies would have to deal with international/transboundary risks. This is what occurred in the Brent Spar case. British civil servants and the government defended Shell's plans to dump the Spar in the deep sea up to the day the company decided not to do it, but this was unlikely to work for several reasons.

First, the British government is not known to be a champion of the environment. It was already known as the 'Dirty Man of Europe' and had been accused of foot-dragging on issues ranging from incinerating waste at sea to cleaning up waterways to reducing long-range air pollution.[63] Second, Shell is not strictly a British company (it is an Anglo/Dutch conglomerate) and the government was therefore only talking to a Shell subsidiary (in this case Shell UK). The decision not to dump Brent Spar was taken by Shell International, the parent company. The directors of Shell International, concerned about the growing effects the consumer boycotts had on Shell subsidiaries in continental Europe (in particular Germany), based their decision on business fundamentals.

Finally, the consensual approach failed because it was not suitable for the problem at hand. Both Shell and Greenpeace are multinational and were therefore 'fighting on several fronts'. Greenpeace exploited this by ensuring that the activists occupying the platform came from different countries, thus resulting in extensive media coverage with the same message: what gives Shell the right to dump trash in the sea? Shell was critically far less flexible; the decision to dump Brent Spar was taken by Shell UK in collaboration with the British government, and not by Shell International in consultation with the European Union and other international and national parties.

Analysis of the risk factors

So what does this case study tell us about the risk factors discussed in Chapter 1? Public trust of the regulator is clearly important. However, public trust towards the other actors is also a key factor in the Brent Spar case.

In a high trust low/high uncertainty risk situation, deliberative risk management strategies are not required. This factor does not apply to this case. The public distrusted the regulators.

In a low public trust situation, risk management strategies are needed but the choice of strategy depends on the reasons for the distrust. In the Brent Spar case the regulator was seen as both partial and incompetent by the public. Both the regulator and the regulatee acknowledged that they did not have the competence required to adequately understand the potential detrimental effects on the marine environment associated with deep sea dumping. As a result, the British Government and Shell brought in scientists (expertise) from the outside to help examine the possible environmental effects. This did not work, however. There was no clear scientific evidence of environmental effects of a dumped Spar on the ocean's fauna and flora. When they acted on incomplete evidence anyway, the public (both British and European) saw the regulators as having only the industry's best interests at heart. As the regulators (or industry for that matter) failed to implement deliberative measures during the prolonged crisis, public distrust of the regulator remained and possibly grew.

Deliberative techniques can help create public trust in a contentious risk management issue, if the public distrust issue has something to do with partiality, but are expensive and time-consuming. Deliberative techniques were not used in this case. The British regulator did not introduce public or stakeholder deliberation to help solve the Brent Spar issue.[64]

In high distrust situations, charismatic individuals can be valuable in negotiating successful deliberative outcomes. The Greenpeace protesters were charismatic individuals. They received most of the press attention, facing Shell's water cannons and resisting Shell's attempts to remove them from the Spar. Neither the government nor Shell had any charismatic individuals to match the protesters.

In any regulatory/risk management process, the political actors, local or national, must be supportive of the final outcome. In the Brent Spar case a consensual style of regulation was implemented. In this case the government found itself defending its actions in the House of Commons, at international environmental meetings, as well as at the G7 summit. At this time the consensual model failed. It failed because, as tradition would have it, government officials avoided discussions with actors outside the government's sphere of influence. In this case Greenpeace was considerably more trusted by the general public than the government, and hence could not be ignored.

It is not enough to assume the regulator or the regulatee has public trust. This must be tested. The DTI mistakenly believed that the public trusted it rather more than Greenpeace. Even when Greenpeace statistics were disproved, public support for their stance remained strong. This suggests high levels of public trust in the group. Had the DTI been more aware of the levels of public trust, they might have taken a more pragmatic approach to communicating with the public on deep sea dumping as the BPEO.

Proactive regulation is more likely to gain public trust. Clearly the DTI was reactive in addressing the controversy around the sinking. This was typical of the British regulatory framework. First they were largely unaware that Greenpeace was campaigning around the sea dumping issues,[65] and second, the regulator had no sense of public opinion about them. The resulting policy vacuum was quickly filled by Greenpeace.[66] Greenpeace, on the other hand, was aware of the strategies that the regulators were proposing. It had lobbied OSPAR negotiations for several years and knew more about what issues concerned the public with regard to offshore dumping.[67]

Interest groups will in many cases try to create public distrust of regulators which in turn can lead to failures of the risk management process. In the Brent Spar case, Greenpeace objectives were to oppose ocean dumping, even though the scientific evidence on the environmental impacts did not fully support this option. Greenpeace got its point across via a massive media campaign with some scientific input to show that the regulators and industry were wrong and they were right: dumping the Brent Spar in the deep sea would have severe environmental repercussions, and there were other viable and more environmentally sound alternatives (dismantling the oil storage buoy on land).

Where regulators are seen as partial, interest groups can help achieve an acceptable risk management outcome. It is clear that in many risk management cases the regulator is not impartial. The regulator was seen by the European public to be promoting industry's best interests in the Brent Spar case. If it had been a purely local issue (that is, just affecting Scotland or even the UK) then the public might have been actively engaged through a consultation process (e.g., citizen panels). But since Brent Spar had an international impact; since the owners of it were multinational corporations; and since most of the opposition to it was from outside the UK (particularly Germany), it was difficult not to involve interest groups, since it was difficult to engage a multitude of publics in a wide array of national/cultural settings.

7
Conclusions: Integrating Trust into Risk Management

The case studies in this book illustrate the importance of the level of public trust in determining the best risk management strategy to employ. This concluding chapter examines the factors that appear most relevant for effective regulation in different contexts and synthesizes them into a decision tree for risk managers. Three major lessons emerge from the case studies as to what leads to effective regulation in various situations, and these will be summarized below.

Deliberation is *not* the be-all and end-all in solving risk management controversies

Although deliberative risk management techniques are currently in vogue – indeed they seem to be perceived by some policy-makers, consultants and industrialists as the *only* effective risk management tool – the preceding case studies suggest this is too simplistic. Deliberation often does have a role to play but the best risk management tool depends on whether, and why, the public does not trust regulators/industry in the first place. As we have seen, if the public regards a regulator as incompetent, a technocratic approach with expert involvement is preferable to involving the public or special interest groups in a deliberative exercise. Similarly, when there is a high level of public trust, policy-makers would benefit from using the technocratic approach of regulation, rather than experimenting with deliberation. These conclusions go against the present trends in regulatory policy.

Trust is inherently a complex topic

Regulators and industry today assume that they are either trusted or not and then act accordingly, sometimes with disastrous results. It is not, as we have seen in this book, sufficient to assume levels of public trust: this needs to be tested. This is not simple, however. Using opinion polls to measure trust is by definition problematic. Qualitative research methods, such as ethnographic surveys, are likely to be much more insightful.

Regulation itself is good

Despite trends towards deregulation and voluntary agreements with industry in a period of declining public trust in regulatory bodies (or at least a perceived decline), it is crucial to reverse this decline to prevent the public losing faith in regulatory agencies completely. This book highlights three reasons for this decline: lack of partiality (fairness), incompetence and inefficiency, or a combination of these factors. Risk managers would do well to focus on these three fundamental reasons for public distrust proactively rather than reacting by finding popular solutions that may only increase public trust in the short term.

The context of the decision-making process

In a high trust but uncertain risk situation, deliberative risk management strategies are not required. The Barsebäck nuclear power plant case shows that given high public trust in the regulatory system, technocratic and rational/economic risk management strategies can function well and deliberative techniques are not necessary. The Swedish regulator implemented its customary technocratic procedure, which was successful. Public trust of both the regulator and industry therefore remained high.

In a low public trust situation, risk management strategies are necessary, but choosing an appropriate strategy will depend largely on the reasons for, if not the context, of the distrust. If the regulator and/or industry is seen by the public as partial, deliberation is needed, as illustrated in the German and US cases studies. In Germany, the North Black Forest respondents did not perceive the local regulators as impartial.

The regulators were forced to develop an intraregional solution for the waste problem by the national government. In other words, they were seen to be as acting on behalf of the state, rather than the local population.

In the USA, IP was not trusted by the local community. It polluted the Androscoggin River and laid off the entire (unionized) workforce. As a private company, IP was also seen as prioritizing shareholder value. The company therefore adopted a deliberative strategy.

Deliberative techniques can help create public trust on a contentious risk management issue, if public distrust stems from a perception of partiality, but are expensive and time-consuming. In Germany and the USA, the two countries in which deliberation risk management strategy was employed, the generation of public trust in the policy-making process was successful. In the North Black Forest three counties agreed on the siting of the waste incinerator and aerobic digesters. Similarly, in Maine, both the IP regulators and the various interest groups agreed on a re-licensing process for the four hydropower stations. In both cases, however, the deliberative process was time consuming and expensive. The deliberative process in the North Black Forest took seven months and cost $1 million, while the re-licensing procedure in Maine took three years and was estimated to cost $4–5 million (most of which was the value of the forest and river frontage donated to the community by IP), but this excluded associated costs to the array of individuals themselves engaged in this process.

Behaviour of risk managers

In high distrust situations, charismatic individuals can be influential in negotiating successful deliberative outcomes. In many cases these individuals can make or break outcomes. Of the four case studies, two had active involvement of charismatic individuals. Brent Spar drew in the dynamic campaigners of Greenpeace, who dominated media coverage for three months. The 'David and Goliath' imagery of these people dodging Shell's speed boats and water cannons captured public attention and sympathy. In the US case, Dan Sosland of the Conservation Law Foundation and Steve Groves of IP, together with other NGOs, State and national regulators and industry representatives, forged the agreement on re-licensing of the hydropower dams. In the Swedish and German cases no individual (or group of individuals) were so conspicuous by comparison.

In any regulatory/risk management process, the political actors – be they local or national – have to be supportive of the final outcome. If this does not occur the final outcome of the risk management process may be worse than if no strategy was put in place. Local and/or national political actors supported the regulatory process in three cases (Sweden, UK and USA). In Sweden, for example, the government supported the Swedish Nuclear Inspectorate's dealings with Sydkraft regarding the blocked filter at the Barsebäck reactor, while in the UK the government (under the then Prime Minister, John Major) supported Shell's decision to dump the buoy in the North Sea. Likewise, in the USA, policy-makers in the state of Maine were highly supportive of IP's deliberative process. Indeed, with regard to both the Swedish and the US cases, the support from the state/national policy makers was pivotal to the success of the risk management process.

The German case, however, is an example of how local policy-makers nearly ruined a successful deliberative process, as they withdrew their support following the completion of the exercise. In the North Black Forest, the main sponsor of the deliberative process, Sigberd Frank, the mayor of Pforzheim, withdrew his support following concern from his fellow CDU party members that allowing an incinerator in Pforzheim would cripple them in the imminent elections.

Public trust cannot be assumed. In these case studies none of the actors involved in the risk management process tested to see if there was 'public trust'. Had they done so at least the UK case could have been better managed. There was considerable debate about siting and building waste incinerators throughout Germany, so the policy-makers in North Black Forest thought a deliberative exercise would help ensure the success of the siting process. They did not feel it necessary to gauge public opinion, however. In the Swedish case, the regulators presumed they had public confidence and therefore did not see the deliberative process as necessary. Had they actually surveyed the public before or after the incident their belief would have been confirmed.[1] IP assumed that the local public did not trust the company, following the firing of the unionized workforce and pollution problems at the plant. Greenpeace's success in halting the sinking demonstrated, moreover, that the public trusted Greenpeace more than the regulator.

Proactive regulation is more likely to gain public trust. In this book three case studies demonstrate proactive regulation: Germany, Sweden and the USA. The UK case represents more of a retroactive process. In Germany, the regulators took the view that public involvement in the policy-making exercise was needed to help ensure passage of the process. Had the risk managers gone ahead and tried to site the waste plants without public involvement, this could have led to public protests and possible failure.

In Sweden, the regulator acted with traditional caution and pro-actively ensured that the risks posed by the filter incident were thoroughly explored. The regulators pushed for this, even though regulators in other nations with similar reactors did not follow suit. By being as forceful as they were, the regulators in Sweden retained the trust and respect of both interest groups and the public, though other nations seem to have largely ignored this example.

With regard to the case in Maine, IP was aware of considerable public distrust and hostility toward the company, and felt a proactive risk management strategy was the best way forward; while in the UK the regulators were decidedly not proactive. First, British decision-makers were supposedly unaware of what Greenpeace was plotting. Second, the regulators did not attempt to gauge public opinion regarding the disposal of offshore oil vessels which could seriously affect the environment.

Special interest groups usually set the public against regulators which often jeopardizes the risk management process. All four case studies involved special interest groups, but the outcomes of their involvement on public trust or distrust varied considerably. In the UK example this hypothesis proved correct. Greenpeace tried to promote the distrust of both the government and Shell and succeeded, which led to a failed regulatory effort.

In Maine there was significant special interest group involvement, such as the Appalachian Mountain Club and the Conservation Law Foundation, but in this case the two main actors from these groups tried to build up public trust through the deliberative process, thereby ensuring a successful risk management process.

The facilitator of the citizen panel process in Germany, the Centre for Technology Assessment, decided early on not to involve interest groups or elected politicians, fearing eventual public distrust of the deliberative process and thus also criticism against themselves. The customary efforts of these special interest actors to undermine the legitimacy of deliberative exercises failed since the public and panellists nevertheless invested their trust in the Centre for Technology.

The Swedish case entailed little interest group involvement. There are several reasons for this. First, the risk managers, in this case the Swedish nuclear inspectorate and Sydkraft, were highly trusted, and hence did not seek out interest group involvement. Second, groups such as Greenpeace simply assumed regulators were indeed acting in the interests of the public rather than themselves, and hence little action was required.

Special interest groups are needed, however, when the regulator is not seen as impartial and when one is dealing with national or international regulatory issues. In many cases the regulator is not impartial. In the UK case of Brent Spar, the regulator was viewed as promoting the industry's best interests. If Brent Spar had been a local case then the public could have been actively involved in some form of policy-making effort. As Brent Spar was an international case, however, with a multi-national industry and multinational stakeholders, involving publics and regulators from all over Europe, a public deliberative process would have been too cumbersome. Although special interest groups in such situations may indeed increase mistrust of the policy-making process, the other options are no better. If special interest groups are not asked to be involved, they can always choose to be regardless (as in the case of Brent Spar), with predictable results concerning trust.

In the Swedish case the regulators were seen as impartial: the Swedish nuclear inspectorate closed down six nuclear reactors against the wishes of the industry. But with the US case and the re-licensing of the hydropower dams on the Androscoggin River, one could argue that this was a local case and hence no special interest group involvement was needed. The re-licensing procedure in Maine was complicated and drawn out, however, with interest groups often meeting for a half a day once a week, lasting up to 3 years. In such cases the general public would be unable to participate in such processes as they would not have the time or resources to do so.

The risk management decision tree

As the implementation of regulatory strategies is inherently nebulous, as seen in the various case studies and summarized by the various

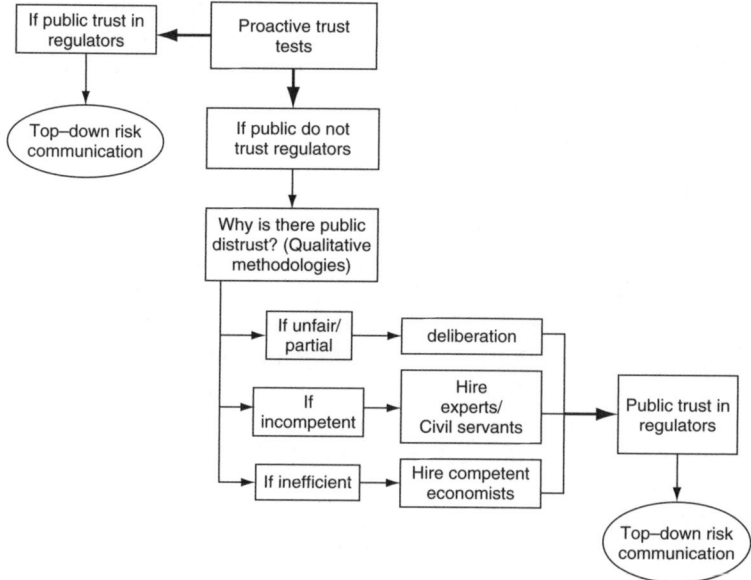

Figure 7.1   Risk management decision tree

risk factors above, how does one go about conducting risk management decisions? A possible tool is a risk management decision tree (Figure 7.1).

If there is trust

As we see from the diagram, if there is trust in the regulator the process can be fairly straightforward: use the regulatory routine that is already in place, as the trust test suggests that the regulator is doing it right. Why change a regulatory design that experience says is working in the promotion of public trust? Yet with every regulatory decision there should be some form of risk communication process established. This would not be a deliberative process in which interest groups and the public are asked to participate actively, but rather more of a top-down form of risk communication, in which the regulator informs the concerned public what is occurring.[2] This is substantially different from encouraging the public or interest groups to *participate* directly in the policy-making process. When there is trust in the policy-making process, the public (and, in particular, special interest groups) should be actively *discouraged* from participating in this process. The principal reasons for this are: first, that the public or special interest groups will wonder why they are suddenly asked to participate; and second, that they might exploit the opportunity to promote their own interests.

If there is distrust

As we see from the diagram, if there is public distrust towards the policy-making process overall, one first has to discern why. The reasons for public distrust should be made clear via the trust test to be implemented prior to the making of a regulatory decision. When the regulator faces a tough decision, knowing himself to be distrusted by the public and why, three possible courses of action emerge.

If the reason for distrust is seen as partiality, a deliberative process is called for as outlined in Chapter 2, in which the public (local issues when it is not complex) or special interest groups (either national/international issues or complex local ones) actively participate in the policy-making process, making suggestions as to what steps ought to be taken. It should be noted, as discussed throughout this book, that the involvement of special interest groups is a high risk strategy. It may increase trust (if the right stakeholder or enlightened individual is involved, as in the US case) or it may lead instead to increased distrust, as was seen in the UK case and Brent Spar. The issue here is that the regulators, driven by a distrustful public, do not have a better alternative. Going to experts or involving economists (as when public distrust is caused by incompetence or inefficiency) is simply not a proper solution, as the public will take the view that the regulator is partial to one group over another.

If the reason for regulator distrust is incompetence, then a technocratic approach is needed in which experts advise civil servants as to what type of regulatory approaches need to be taken. Such an approach will work if the public believes that the experts are impartial and competent. It can, however, fail if there is no scientifically agreed consensus regarding the issue at hand, as scientific infighting can also lead to public distrust. In order to ensure impartiality when confronted by scientific pluralism, a blue ribbon set of impartial scientists needs to be formed, as was the case with Brent Spar where the Sheppard Commission was established to resolve several conflicting scientific claims.

If the reason for public distrust is that the regulator is perceived as inefficient, then an economic/rational risk approach is needed. Such an approach will involve a number of economists both within and outside the government, focusing on how best to use the limited amount of funding available for the regulatory process. In making the process more efficient, these economists will examine market-oriented systems such as tradable pollution rights, pollution taxes, and monetary incentives for pollution prevention. An example of the rational risk management being used is the US regulatory approach of the post-1980 era with the involvement of OMB, an attempt to make regulators more accountable for the costs as well as benefits imposed by regulations. This process will work only if the public believes economists increase the efficiency of the overall process, and are not simply a waste of tax payers' money. Any indication of partiality on the part of the economists themselves will also undermine the process.

If there is more than one reason why the public distrusts the regulatory process, as was the case in the UK where regulators were seen to be both partial *and* incompetent, then some form of regulatory reform is needed in order to re-establish public trust. These reforms can either be radical, as in the case of the Netherlands (where rampant public distrust towards policy-makers made public deliberation a virtual modus operandi for more than a decade[3]), or less drastic, albeit still significant, such as the recent establishment of the UK Environment Agency following years of criticism of Her Majesty's Inspectorate of Pollution (HMIP), commonly regarded as 'industry-friendly'.[4]

In a best case scenario, however, the regulator should not have to implement any of these options to reduce public distrust. It should have been carrying out these so-called 'trust tests' proactively, implementing some form of trust-enhancement measures far in advance of any anticipated crisis. If the regulator is seen as being partial, for example, it would act proactively before an incident becomes a reality by working with a wide array of actors to ensure that the regulator is working in everyone's best interests. In so doing some form of consensus will be reached, leading to public trust in regulators.

Final words

The lessons from this book apply to local, national, multinational and global regulatory bodies. Although in Europe, increasingly more regulations are being dealt with at the EU Commissioner level, the risk management factors that came out of the four cases also apply to the Commission. The issue here is one mainly of scale, going from national to multinational. In any local, national or multinational case the following can, for example, be observed:

1 In a high public trust environment, extensive direct public deliberation will not be needed.
2 In a low public trust environment, deliberation in the form of special interest groups or publics, technocrats in the form of experts or scientists, or economists is necessary. The extent of their involvement in the policy-making process depends on the reasons for public distrust.
3 Involving interest groups in the policy-making process because of impartiality is a risky process, and may increase public distrust.

In sum, it is hoped that this book has provided some guidance on helping risk managers, be they local, national or international, to solve the ongoing problem of growing public distrust of the regulatory process. *Regulation is good* since it offers advantages for efficiency *and* equity, carrying many obvious benefits for public and environmental prosperity.

Notes and References

Preface

1. Anthony Giddens, *The Consequences of Modernity* (Cambridge: Polity Press, 1990); Joseph S. Nye Jr, Philip D. Zelikow and David C. King, *Why People Don't Trust Government* (Cambridge, MA: Harvard University Press, 1997).
2. Ragnar E. Löfstedt and David Vogel, 'The changing character of regulation: a comparison between Europe and the United States', *Risk Analysis*, 21(3), 2001, 399–405.
3. Corie Lok and Douglas Powell, *The Belgian Dioxin Crisis of the Summer of 1999: A Case Study in Crisis Communication and Management* (Guelph, Ontario: Department of Food Science, University of Guelph, 2000).
4. Scott Ratzan (ed.) *The Mad Cow Crisis: Health and the Public Good* (London: University College London Press, 1999).
5. K. Newton and Pippa Norris, 'Confidence in public institutions: fear, culture or performance?', in S. Pharr and Robert Putnam (eds), *Disaffected Democracies: What's Troubling the Trilateral Countries?* (Princeton: Princeton University Press, 2000).
6. House of Lords, *Select Committee on Science and Technology: Science and Society* (London: Stationary Office, 2000)
7. Soren Holmberg and Lennart Weibull, (eds), *Ljusnande Framtid* (Gothenburg: University of Gothenburg, SOM Institute, 1999)
8. Paul Slovic, 'Perceived risk, trust and democracy', *Risk Analysis*, 13, pp. 675–82.
9. Ragnar Löfstedt, 'Risk communication: the Barseback nuclear plant case', *Energy Policy*, 24(8), 689–96.
10. Timothy Earle and George Cvetkovich, *Social Trust: Toward a Cosmopolitan Society* (Westport CT: Praeger, 1995).
11. George Cvetkovich and Ragnar E. Löfstedt, *Social Trust and the Management of Risk* (London: Earthscan, 1999).
12. Ortwin Renn and Deborah Levine, 'Credibility and trust in risk communication', in Roger E. Kasperson and Peter Jan Stallen (eds), *Communicating Risks to the Public: International Perspectives* (Amsterdam: Kluwer, 1991).
13. Ortwin Renn, 'Die Austragung offentlicher Konflikte und chemische Produkte oder Produktionsverfahren-eine soziologische Analyse', in Ortwin Renn and Jurgen Hampel (eds), *Kommunikation und Konflikt: Fallbeispiele aus der Chemie* (Wurzburg: Konigshausen und Neumann, 1996).

1   Introduction and Overview

1. For a more detailed discussion regarding this example see Ragnar E. Löfstedt, 'Risk and regulation: boat owners' perceptions to recent anti-fouling legislation', *Risk Management an International Journal*, 3:3 (2001), 33–46.

2. See for example, House of Lords, Select Committee on Science and Technology, *Science and Society* (London: The Stationery Office, 2001); National Research Council (NRC), *Improving Risk Communication* (Washington, DC: National Academy Press, 1989); NRC, *Understanding Risk* (Washington, DC: National Academy Press, 1996); Royal Commission for Environmental Pollution (RCEP), *Setting Environmental Standards* (London: The Stationery Office, 1998); Strategy Unit, *Risk: Improving Government's Capability to Handle Risk and Uncertainty* (London: Strategy Unit, Cabinet Office, 2002).

3. For a good discussion of this see Susan D. Pharr and Robert D. Putnam (eds), *Disaffected Democracies: What's Troubling the Trilateral Countries?* (Princeton, NJ: Princeton University Press, 2000).

4. See, for example, Ronald Inglehart, *Culture Shift in Advanced Industrial Society* (Princeton, NJ: Princeton University Press, 1990); Sheila Jasanoff, *The Fifth Branch: Science Advisors as Policy Makers* (Cambridge, MA: Harvard University Press,1990); Ragnar E. Löfstedt and David Vogel, 'The changing character of regulation: A comparison of Europe and the United States', *Risk Analysis*, 21:3 (2001), 399–405, Joseph S. Nye Jr, P.D. Zelikow and D.C. King, *Why People Don't Trust Government* (Cambridge, MA: Harvard University Press, 1997); and Robert D. Putnam, *Bowling Alone* (New York: Basic Books, 2000).

5. 'Risk' is, for the sake of argument, defined here more broadly as hazard rather than the probability of a hazard. A risk manager is someone who manages either an actual or a perceived (socially constructed) hazard. Examples of risk managers would be state or national regulators (such as the Swedish Chemical Inspectorate or the US Environmental Protection Agency), as well as industrial health, safety and environment representatives. In this book I focus primarily on the public's trust of local, state or national risk managers, more commonly referred to as the public regulators.

6. Paul Slovic, 'Perceived risk, trust and democracy', *Risk Analysis*, 13:6 (1993), 675–82.

7. Anthony Giddens, *The Consequences of Modernity* (Cambridge: Polity Press, 1990).

8. Stephen Breyer, *Breaking the Vicious Circle: Toward Effective Risk Regulation* (Cambridge, MA: Harvard University Press, 1993); C. Coglianese, 'The limits of consensus: the environmental protection system in transition: toward a more desirable future', *Environment*, 41:3 (1999), 1–6.

9. NRC, *Understanding Risk*; RCEP, *Setting Environmental Standards*; Ortwin Renn, 'A model for an analytic-deliberative process in risk management', *Environmental Science and Technology*, 33:18 (1999), 3,049–55.

10. Robert A. Kagan and Lee Axelrad, *Regulatory Encounters: Multinational Corporations and American Adversarial Legalism* (Berkeley, CA: University of California Press, 2000); Steven Kelman, *Regulating America, Regulating Sweden: A Comparative Study of Occupational Safety and Health Policy* (Cambridge, MA: MIT Press, 1981); Leif Lewin, *Bråka inte! Om vår tids demokratisyn* (*Do not fight: About our time's view on democracy*) (Stockholm: SNS Forlag, 1998); David Vogel, *National Styles of Regulation: Environmental Policy in Great Britain and the United States* (Ithaca, NY: Cornell University Press, 1986).

11. Carnegie Commission on Science, Technology and Government, *Risk and the Environment: Improving Regulatory Decision Making* (Washington, DC: Carnegie Commission, 1993).

12. Breyer, *Closing the Vicious Circle*; William D. Ruckelshaus, 'Science, risk and public policy', *Science*, 221 (1983) 1,026–8; William D. Ruckelshaus, 'Risk, science, and democracy', *Issues in Science and Technology*, 1:3 (1985), 19–38.

13. Sören Holmberg and Lennart Weibull (eds), *Ett missnöjt folk?* (*An unhappy people?*) (Gothenburg: SOM Institute, Department of Political Science, University of Gothenburg, 1997); House of Lords, *Science and Society*; Pharr and Putnam, *Disaffected Democracies*. This is particularly strong in Europe lately due to food safety scares such as BSE ('mad cow disease') and foot and mouth disease.

14. Richard H. Pildes and Cass R. Sunstein, 'Reinventing the regulatory state', *University of Chicago Law Review*, 62:1 (1995), 1–129; RCEP, *Setting Environmental Standards*.

15. In this book interest groups are synonymous with non-governmental organizations (NGOs) such as Greenpeace. Although the public may also be seen as an interest group in many cases, for the sake of argument I do not include the public as an interest group (see Chapter 2, the section on deliberation, for more of a discussion).

16. Roderic M. Kramer and Tom R. Tyler, *Trust in Organizations: Frontiers of Theory and Research* (Thousand Oaks, CA: Sage, 1996).

17. Anthony Giddens, *Consequences of Modernity*, p. 34.

18. Timothy Earle and George Cvetkovich, *Social Trust: Toward a Cosmopolitan Society* (Westport, CT: Praeger, 1995).

19. Ortwin Renn and Deborah Levine, 'Credibility and trust in risk communication', in Roger E. Kasperson and Peter Jan Stallen (eds), *Communicating Risks to the Public: International Perspectives* (Amsterdam: Kluwer, 1991).

20. Slovic, 'Perceived risk, trust and democracy'.

21. Robert W. Hahn, (ed.), *Risks, Costs and Lives Saved: Getting Better Results from Regulation* (New York: Oxford University Press, 1996); W. Kip Viscusi, *Rational Risk Policy* (New York: Oxford University Press, 1998).

22. For a good review see Paul Slovic, 'Perception of risk', *Science*, 236 (1987), 280–5.

23. R. Brickman, S. Jasanoff and T. Ilgen, 'Controlling Chemicals: The politics of regulation in Europe and the united states' (Ithaca, Cornell University Press 1985); Kelman, *Regulating Sweden, Regulating United States*.

24. Kelman, *Regulating Sweden, Regulating United States*.
25. Ragnar E. Löfstedt, 'Risk communication: The Barsebäck nuclear plant case', *Energy Policy*, 24:8 (1996), 689–96; Slovic, 'Perceived risk, trust and democracy'. It should be noted that these findings have been challenged by Lennart Sjöberg in his article, 'Perceived competence and motivation in industry and government as factors in risk perception', in George Cvetkovich and Ragnar E. Löfstedt (eds), *Social Trust and the Management of Risk* (London: Earthscan, 1999).
26. For an excellent discussion on this topic see Aaron Wildavsky, *Searching for Safety* (New Brunswick, NJ: Transaction Books, 1988).
27. See, for example, Chris Hohenemser, Roger E. Kasperson and Rober W. Kates, 'The distrust of nuclear power', *Science*, 196 (1977), 25–34.
28. NRC, *Understanding Risk*; RCEP, *Setting Environmental Standards*. For good overviews on the findings in the trust literature please consult George Cvetkovich and Ragnar E. Löfstedt (eds), *Social Trust and the Management of Risk* (London: Earthscan,1999); Earle and Cvetkovich, *Social Trust: Toward a Cosmopolitan Society*; Diego *Gambetta, Trust: Making and Breaking Cooperative Relations* (Oxford: Basil Blackwell, 1988); Roger E. Kasperson, Dominic Golding and Seth Tuler, 'Siting hazardous facilities and communicating risks under conditions of high social distrust', *Journal of Social Issues*, 48 (1992), 161–72; and Barbara A. Misztal, *Trust in Modern Societies* (Cambridge: Polity Press, 1996).
29. See for example, Richard J. Lazarus, 'The tragedy of distrust in the implementation of federal environmental law', *Law and Contemporary Problems*, 54 (1991), 311–74.
30. Douglas Powell and William Leiss, *Mad Cows and Mother's Milk: The Perils of Poor Risk Communication* (Montreal: McGill-Queen's University Press, 1997).
31. An example of this was the recent 'Ghost ship debacle' in the UK. For a detailed discussion see Ragnar E. Löfstedt, *How can better Risk Management lead to greater Public Trust in Canadian Institutions: Some Sobering lessons from Europe* (London: King's Centre for Risk Management, 2004).
32. See, for example, Ragnar E. Löfstedt, 'Evaluation of two siting strategies: the case of two UK waste tyre incinerators', *Risk: Health Safety and Environment*, 8:1 (1997), 63–77.

2   A Review of the Four Risk Management Strategies

1. M. Weber, *Basic Concepts in Sociology* (New York: Philosophical Library, 1962).
2. T.E. Parsons, *Sociological Theory and Modern Society* (New York: Free Press, 1967).
3. O. Renn, 'Die Austragung offentlicher Konflikte um Chemische Produkte oder Produktionsverfahren-eine soziologische Analyse' ('The conducting of public conflicts in the chemical area or the production procedure-an analysis') in O. Renn and J. Hampel (eds), *Kommunikation und Konflikt:*

Fallbeispiele aus der Chemie (*Communication and conflict: Examples from the chemical area*) (Wurzburg: Konigshausen & Neumann, 1996).

4. M.K. Landy, M.J. Roberts and S.R. Thomas, *The Environmental Protection Agency. Ask the Wrong Questions from Nixon to Clinton* (expanded edition) (New York: University Press, 1994).

5. House of Lords. Science and Technology Committee, *Science in Society* (London: House of Lords, 2000).

6. J.D. Graham and J.B. Wiener, *Risk vs. Risk* (Cambridge, MA: Harvard University Press, 1995).

7. J. Ramsberg, *Are All Lives of Equal Value: Studies of the Economics of Risk Regulation* (Stockholm: Stockholm School for Economics, 1999).

8. S. Breyer, *Breaking the Vicious Circle: Toward Effective Risk Regulation* (Cambridge, MA: Harvard University Press, 1993); W.K. Viscusi, *Rational Risk Policy: The 1996 Arne Ryde Memorial Lectures* (New York: Oxford University Press, 1998).

9. D. Fahrni, *An Outline History of Switzerland: From the Origins to the Present Day* (Zurich: Pro Helvetia Arts Council, 1992).

10. O. Renn, T. Webler and R.E. Löfstedt, *The Challenge of Integrating Deliberation and Expertise: Models of Participation and Discourse in Risk Management* (Stuttgart: Centre for Technology Assessment, 2000).

11. Ibid.

12. T. Lowi, *The End of Liberalism: The Second Republic of the United States* (New York: W.W. Norton, 1979).

13. Ibid.

14. D. Fiorino, 'Environmental risk and democratic process: A critical review'. *Columbia Journal of Environmental Law*, Vol. 14 (1989), 501–47.

15. Ibid., p. 504.

16. P.C. Dienel, *Die Planungzelle* (Opladen: Westdeutscher Verlag, 1978); O. Renn, 'A model for an analytic-deliberative process in risk management', *Environmental Science and Technology*, 33:18, 3,049–55.

17. See, for example, RCEP, *Setting Environmental Standards* (London: The Stationery Office, 1998).

18. C. Chess and K. Purcell, 'Public participation and the environment: Do we know what works?', *Environmental Science and Technology*, 33:16, 2,685–92; O. Renn, T. Webler and P. Wiedemann, *Fairness and Competence in Citizen Participation* (Kluwer: Dordrecht, 1995).

19. NRC, *Understanding Risk* (Washington, DC: National Academy Press, 1996).

20. N. Pidgeon, 'Stakeholders, decisions and risk', in A. Mosleh and R.A. Bari (eds), *Probabilistic Safety Assessment and Management, PSAM 4*, 3 (1997) 1,583–8; J. Rossi, 'Participation run amok: The costs of mass participation for deliberative agency decisionmaking', *Northwestern University Law Review*, 92:1 (1997), 173–250.

21. B.R. Barber, *Strong Democracy: Participatory Politics for a New Age* (Berkeley, CA: University of California Press, 1984).

22. W.E. Wagner, 'The science charade in toxic risk regulation', *Columbia Law Review*, 95:77, 1,613–723 (1995) Brooks cited on p. 46.

23. J. Madison, 1787, 'The union as a safeguard against domestic faction and insurrection'. Federalist Paper 10, New York.
24. M. Watts, *Silent Violence* Berkeley (University of California Press, 1983).
25. K. Shrader-Frechette, 'Scientific method, anti-foundationalism, and public policy', *Risk: Issues in Health and Safety*, 1 (1990), 23–41.
26. P. Slovic, 'Perception of risk', *Science*, 236 (1993), 280–85; B. Wynne, 'Sheepfarming after Chernobyl: a case study in communicating scientific information', *Environment*, 31:2 (1989), 10–15, 33–9; B. Wynne, 'May the sheep safely graze? A reflexive view of the expert–lay knowledge divide', in S. Lash, B. Szerszynski and B. Wynne (eds), *Risk, Environment and Modernity: Towards a New Ecology* (London: Sage, 1996).
27. B. Fischhoff, 'Risk perception and communication unplugged: Twenty years of process', *Risk Analysis*, 15 (1995), 137–45; W. Leiss, 'Three phases in the evolution of risk communication practice', *Annals of the American Academy of Political and Social Science*, 545 (1996), 85–94; NRC, *Improving Risk Communication* (Washington, DC: National Academy Press, 1989).
28. R. Adler and D. Pittle, 'Cajolery and command: are education campaigns an adequate substitute for regulation? *Yale Journal on Regulation*, 2 (1984), 159–94; P. Slovic and D. MacGregor, *The Social Context of Risk Perception*, (Decision Research, Eugene Oregon; 1984).
29. E. Siddall and C.R. Bennett, 'A people-centered concept on society-wide risk management', in R.S. McColl (ed.), *Environmental Health Risks: Assessment and Management* (Waterloo, Ontario: University of Waterloo Press, 1987).
30. Leiss, 'Three phases'.
31. Rossi, 'Participation run amok'.
32. NRC, *Understanding Risk*; Presidential/Congressional Commission on Risk Assessment and Risk Management, *Risk Assessment and Risk Management in Regulatory Decision-Making*, Final report (Washington, DC: 1997); Renn, Webler and Wiedemann, *Fairness and Competence in Citizen Participation*; RCEP, *Setting Environmental Standards*.
33. Renn, Webler and Wiedemann, *Fairness and Competence in Citizen Participation*.
34. House of Lords, Science and Technology Select Committee, *Science in Society* (London: House of Lords, 2000).
35. S. Jasanoff, *Science at the Bar: Law, Science and Technology in America* (Cambridge, MA: Harvard University Press, 1995); Wynne, 'May the sheep safely graze?'.
36. H. Brooks, 'The resolution of technically intensive public policy disputes', *Science, Technology and Human Values*, Winter (1984) 39–; M.G. Kweit and R.W. Kweit, 'The politics of policy analysis: the role of citizen participation in analytic decision making', in J. DeSario and S. Langton (eds), *Citizen Participation in Public Decision Making* (1987).
37. R.J. Rydell, 'Solving political problems of nuclear technology: the role of public participation', in DeSario and Langton, *Citizen Participation in Public Decision Making*.

38. Wynne, 'Sheepfarming after Chernobyl'; Wynne, 'May the sheep safely graze?'.
39. For a review of the various models of participatory deliberation please see Renn, Webler and Löfstedt, *The Challenge of Integrating Deliberation and Expertise.*
40. Rossi, 'Participation run amok'.
41. Ibid.
42. S.A. Shapiro, 'Political oversight and the deterioration of regulatory policy', *Administration Law Review*, 46 (1994).
43. E. Aronson, *The Social Animal* (San Francisco: W.H. Freeman, 1999).
44. Breyer, *Breaking the Vicious Circle*; S. Jasanoff, *The Fifth Branch: Science Advisors as Policymakers* (Cambridge, MA: Harvard University Press, 1990); D. Vogel, *National Styles of Regulation: Environmental Policy in Great Britain and the United States* (Ithaca, NY: Cornell University Press, 1986).
45. Vogel, *National Styles of Regulation.*
46. Jasanoff, *The Fifth Branch*, ch. 2.
47. Jasanoff, *Science at the Bar.*
48. This is sometimes known as expertocratic decision-making.
49. R. Pildes and C.R. Sunstein, 'Reinventing the regulatory state', *University of Chicago Law Review*, 62:1 (1995), 1–129.
50. Breyer, *Breaking the Vicious Circle.*
51. Ibid.
52. Paraphrased from J.D. Graham and J.K. Hartwell, 'The risk management approach', in J.D. Graham and J.K. Hartwell (eds), *The Greening of Industry: A Risk Management Approach* (Cambridge, MA: Harvard University Press, 1997), pp. 1–2.
53. Breyer, *Breaking the Vicious Circle*; F.B. Cross, *Legal Responses to Indoor Air Pollution* (New York: Quorum Books, 1990); F.B. Cross, 'The public in risk control', *Environmental Law*, 24 (1994), 888–969.
54. Brickman, Jasanoff and Ilgen, *Controlling Chemicals*; Federal Focus Inc., *Toward Common Measures: Recommendations for a Presidential Executive Order in Environmental Risk Assessment and Risk Management Policy* (Washington, DC: Federal Focus Inc., 1991).
55. J.D. Graham and J.K. Hartwell (eds), *The Greening of Industry: A Risk Management Approach* (Cambridge, MA: Harvard University Press, 1997); NRC, *Risk Assessment in the Federal Government: Managing the Process* (Washington, DC: National Academy Press, 1983); W. Ruckelshaus, 'Science, risk and public policy', *Science*, 221 (1983), 1,026–8. W. Ruckelshaus, 'Risk, science and democracy', *Issues in Science and Technology*, 1:3 (1985), 19–38.
56. Graham and Wiener, *Risk vs. Risk.*
57. C. Anderson, 'Cholera epidemic tied to risk miscalculation', *Nature*, 354 (28 November 1991), 255.
58. Vogel, *National Styles of Regulation.*
59. S. Kelman, *Regulating America, Regulating Sweden: A Comparative Study of Occupational Safety and Health Policy* (Cambridge, MA: MIT Press, 1981).

60. Ruckelshaus, 'Science, risk and public policy'; Ruckelshaus, 'Risk, science and democracy'.

61. A. Gore, *From Red Tape to Results: Creating a Government that Works Better and Costs Less, Report of the National Performance Review* (Washington: GPO, 1993); D. Osborne and T. Gebler, *Reinventing Government: How the Entrepeneurial Spirit is Transforming the Public Sector* (Boston, MA: Addison-Wesley, 1991).

62. Vogel, *National Styles of Regulation*.

63. Pildes and Sunstein, 'Reinventing the regulatory state'.

64. D.A. Dana, 'Review essay: setting environmental priorities: the promise of a bureaucratic solution: breaking the vicious circle: toward effective risk regulation', *Boston University Law Review*, 74 (1994), 365–; R.A. Pollak, 'Regulating risks', *Journal of Economic Literature*, Vol. 33 (1995), 179–91.

65. T. McGarity, 'Substantive and procedural discretion in administrative resolution of science policy questions: Regulating carcinogens in EPA and OSHA', *Georgetown Law Journal*, 67 (1979), 729–.

66. D.A. Wirth and E.K. Silbergeld, 'Risk reform', *Columbia Law Review*, 95 (1995), 1,857–95.

67. N. Ashford *et al.*, 'A hard look at federal regulation of formaldehyde: a departure from reasoned decisionmaking', *Harvard Environmental Law Review*, 7 (1983), 297–.

68. S.E. Gaines, 'Science, politics and the management of toxic risks through law', *Jurimetrics Journal*, 30 (1990), 271–.

69. W.E. Wagner, 'The science charade in toxic risk regulation', *Columbia Law Review*, 95:77 (1995), 1,613–723.

70. W. Freudenburg, 'Perceived risk, real risk: social science and the art of probabilistic risk assessment', *Science*, 242 (1988), 44–9.

71. L. Heinzerling, 'Regulatory costs of mythic proportions', *Yale Law Journal*, 107 (1988), 1,981–; L. Heinzerling, 'Clean air and the constitution', *St Louis University Public Law Review*, 20 (2001), 151–. F. Ackerman and L. Heinzerling, *Priceless: On knowing the price of everything and the value of nothing* (New York, NY: The New Press, 2004).

72. Paraphrased from J.D. Graham, 'The risk management approach', in J. Graham and K. Hartwell (eds), *The Greening of Industry: A Risk Management Approach* (Cambridge, MA: Harvard University Press, 1997), 3–4.

73. J.D. Graham (ed.), *Preventing Automobile Injury: New Findings from Evaluation Research* (Dover, MA: Auburn House, 1988).

74. W.K. Viscusi, J.M. Vernon and J. Harrington Jr, *Economics of Regulation and Antitrust* (Cambridge, MA: MIT Press, 1995), 24.

75. Ibid., p. 26.

76. For an excellent discussion on the advent of rational risk policy in the USA, see ibid., pp. 22–7.

77. RCEP, *Setting Environmental Standards*.

78. For a detailed discussion see R.E. Löfstedt, 'The swing of the regulatory pendulum in Europe: from precautionary principle to (regulatory) impact analysis', in *Journal of Risk and Uncertainty*, 28:3 (2004), 237–60.

79. H.C. Kunreuther, R. Ginsberg, L. Miller, P. Sagi, P. Slovic, B. Borkin and N. Katz, *Disaster Insurance Protection: Public Policy Lessons* (New York: Wiley, 1978).
80. See the following, for example: R. Hahn, R. Lutter and W.K. Viscusi, *Do Federal Regulations Reduce Mortality?* (Washington, DC: American Enterprise Institute, 2000); C. Sunstein, *Risk and Reason: Safety, Law and the Environment* (New York: Cambridge University Press, 2002); W.K. Viscusi, *Fatal Tradeoffs: Public and Private Responsibilities to Risk* (New York: Oxford University Press, 1992); W.K. Viscusi, *Rational Risk Policy: The 1996 Arne Ryde Memorial Lectures* (New York: Oxford University Press, 1998); A.L. Nichols and R.J. Zeckhauser, 'The perils of prudence: how conservative risk assessments distort regulation', *Regulation*, November/December (1986), 13–24; R.J. Zeckhauser, 'Procedures for valuing lives', *Public Policy*, 23:4 (1975), 419–64.
81. Dana, 'Review essay', p. 366.
82. Breyer, pp. 11–12, taken from Viscusi, *Rational Risk Policy*, p. 93.
83. A. Tversky and D. Kahneman, 'Judgement under certainty: heuristics and biases', *Science*, 185 (1974), 1,124–31.
84. For example, B. Fischhoff, P. Slovic, S. Lichtenstein, S. Read and B. Combs, 'How safe is safe enough? A psychometric study of attitudes toward technological risks and benefits', *Policy Studies*, 9 (1978), 127–52.
85. See also C. Sunstein, *Risk and Reason: Safety, Law and the Environment* (New York: Cambridge University Press, 2002).
86. Viscusi, *Rational Risk Policy*.
87. Carnegie Commission on Science, *Technology and Government, Risk and the Environment: Improving Regulatory Decisionmaking* (Washington, DC: Carnegie Commission, 1995); Nichols and Zeckhauser, 'The perils of prudence'; Sunstein, *Risk and Reason*.
88. Viscusi, *Rational Risk Policy*.
89. Ibid., 99–100.
90. Ibid. *Rational Risk Policy*.
91. Hahn, Lutter and Viscusi, *Do Federal Regulations Reduce Mortality?*; T.O. Tengs, M.E. Adams, J.S. Pliskin, D.G. Safran, J.E. Siegel, M.C. Weinstein and J.D. Graham, 'Five-hundred life-saving interventions and their cost effectiveness', *Risk Analysis*, vol. 13 (1995), 369–90.
92. K. Schneider, 'New view calls environmental policy misguided', *New York Times*, 21 March (1993), 1.
93. Shrader-Frechette, 'Scientific method, anti-foundationalism, and public policy'.
94. L. Heinzerling, 'Political Science', *University of Chicago Law Review*, 62 (1995), 449–73; Heinzerling, 'Clean air and the constitution'.
95. Heinzerling, 'Regulatory costs of mythic proportions'.
96. C. Sunstein, 'Democratizing America through law', *Suffolk University Law Review*, 24 (1991), 949–80; Sunstein, *Risk and Reason*.
97. E.K. Silbergeld, 'Risk assessment and risk management: an uneasy divorce', in D.G. Mayo and R.D. Hommander (eds), *Acceptable Evidence: Science and Values in Risk Management* (New York: Oxford University Press, 1991).

98. Pildes and Sunstein, 'Reinventing the regulatory state'.
99. L.H. Tribe, 'Policy science: Analysis or ideology?', *Philosophy and Public Affairs*, 2 (1972), 66–.
100. For example, J. Adams, *Risk* (London: University College London Press, 1995); B. Fischhoff, 'Heuristics and biases in application', in T. Gilovich *et al.* (eds), *Heuristics and Biases: The Psychology of Intuitive Judgement* (New York: Cambridge University Press, 2002).

3   Germany and the Waste Incinerator in the North Black Forest

1. Peter Katzenstein, *Policy and Politics in Western Germany: The Growth of a Semi-sovereign State* (Philadelphia, PA: Temple University Press, 1987); Peter Katzenstein, 'The Third West German Republic: continuity in change', *International Journal of Foreign Affairs* (1998), 325–44; Richard Munch, 'The political regulation of technological risks', *International Journal of Comparative Sociology*, 36 (1995), 109–30.
2. Ortwin Renn, Thomas Webler and Peter Wiedemann (eds), *Fairness and Competence in Citizen Participation* (Dordrecht: Kluwer, 1995).
3. R. Dahrendorf, *Gesellschaft und Demokratie in Deutschland* (*Society and democracy in Germany*) (Munich: DTV, 1968).
4. E. Hartrich, *The Fourth and Richest Reich: How the Germans Conquered the Postwar World* (New York: Macmillan, 1980); Katzenstein, *Policy and Politics in Western Germany*.
5. Dieter Lorenz, 'The constitutional supervision of the administrative agencies in the Federal Republic of Germany', *Southern California Law Review*, 53: 2, (1980), 543–82.
6. D.P. Currie, 'Air pollution control in Western Germany', *University of Chicago Law Review*, 49: 2 (1982), 359–60.
7. See Helmut Wiedner, 'Environmental policy and politics in Germany', in Uday Desai (ed.), *Environmental Politics and Policy in Industrialized Countries* (Cambridge, MA: MIT Press, 2002).
8. Robert Coppock, *Regulating Chemical Hazards in Japan, West Germany, France, the United Kingdom, and the European Community: A Comparative Examination* (Washington, DC: National Academy Press, 1986).
9. Munch, 'The political regulation of technological risks', p. 112.
10. C. Hey and U. Brendle, *Umweltverbande und EG. Strategien, Politische Kulturen und Organisationsformen* (*Environmental groups and the EU: Strategies, Political Cultures and Forms of Organisation*) (Opladen: Westdeutscher Verlag, 1994).
11. Susan J. Pharr and Robert D. Putnam (eds), *Disaffected Democracies: What's Troubling the Trilateral Countries?* (Princeton, NJ: Princeton University Press, 2000).
12. Sonja Boehmer-Christiansen and Jim Skea, *Acid Politics* (London: Belhaven, 1991).
13. See R. Koopmans, *Democracy from Below: New Social Movements and the Political System in West Germany* (Boulder, CO: Westview Press 1995).

14. David Vogel, *Trading Up: Consumer and Environmental Regulation in a Global Economy* (Cambridge, MA: Harvard University Press, 1995).
15. Heinrich Pehle, 'Germany: Domestic obstacles to an international forerunner', in Mikael Skou Andersen and Duncan Liefferink (eds), *European Environmental Policy: The Pioneers* (Manchester: Manchester University Press, 1997).
16. Heinrich Pehle and Alf Inge Jansen, 'Germany: the engine in European environmental policy?', in Kenneth Hanf and Alf Inge Jansen (eds), *Governance and Environment in Western Europe* (Harlow: Longman, 1998).
17. Munch, 'The political regulation of technological risks', pp. 112–14.
18. This case description is adopted from R. Löfstedt, 'The role of trust in the North Black Forest: an evaluation of a citizen panel project', *Risk: Health Safety and Environment*, 7 (1999), 10–30. The sections from this article have been reprinted with permission.
19. Ortwin Renn, Thomas Webler and Hans Kastenholz, 'Procedural and substantive fairness in landfill siting', *Risk: Health, Safety and Environment*, 7 (1996), 145–68.
20. *Südwestpresse*, 'Drei Grunde gegen die Mullverbrennung' ('Three reasons against waste incineration'), 22 November 1995.
21. *Schwarzwalder Bote*, 'Im Burgerforum werden Fakten vermisst' ('In the Citizen Advisory Boards Facts become missed'), 16 January 1996.
22. *Schwarzwalder Bote*, 'Mullmenge hat sich in funf Jahren halbiert' ('Levels of trash have halved over the past five years'), 26 July 1996.
23. For example, the waste could be incinerated in other towns such as Tubingen where there would be excess capacity due to recycling initiatives, or the waste could be landfilled locally.
24. Mayors Theurer and Frank were chosen as they represent opposing views in the discussions regarding the usefulness of citizen juries in general. In the content analysis, Frank was portrayed as the most positive of the policy-makers about the jury concept, and Theurer one of the most negative. In this regard they do not fully represent the views of all the policy-makers in the Black Forest region. In fact, the mayors of the other towns in the area were also somewhat sceptical of the process.
25. Peter C. Dienel, 'Partizipation an Planungsprozessen als Aufgabe der Verwaltung' ('Participation in planning processes as an exercise of administration'), *Die Verwaltung* (1971), 151; Peter C. Dienel and Ortwin Renn, 'Planning cells: a gate to "fractal" mediation', in Renn, Webler and Wiedemann, *Fairness and Competence in Citizen Participation*.
26. The interviews with the panelists were conducted in German and taped for further analysis while the interviews with Frank and Theurer were not taped and the quotations from them are based on shorthand notes. As the quotations here have been translated they are not verbatim, but simply give an accurate overall description of what the interviewee said.
27. Ortwin Renn, 'Premises of risk communication: results of two participatory experiments', in Roger E. Kasperson and Peter J. Stallen (eds), *Communicating Risks to the Public: International Perspectives* (Dordrecht: Kluwer, 1991).

28. Johannes Klomfass, 'Ausserdem-so ein Schwindel!' ('Besides such as swindle!') *Sudwest Presse*, 11 January 1996.
29. *Schwarzwalder Bote*, 'Im Burgerforum werden Fakten vermisst' ('In the citizen advisory boards facts become missed'), 16 January 1996.
30. *Schwarzwalder Bote*, Am Rande Bemerkt: Fingerzeig, 20 January 1996. This accusation is not true. In fact, the citizen panel concept requires the panels to be made up of individuals from different parts of the area in question See Dienel and Renn, 'Planning Cells'; Hans Jorg Seiler, 'Review of "Planning Cells": A Problem of Legitimation', in Renn, Webler and Wiedemann, *Fairness and Competence in Citizen Participation*.
31. *Sudwest Presse*, 'Skandaloses Scheinverfahren' ('Scandalous fictitious act'), 25 January 1996.
32. Harald Friedrich, 'Versuchskaninchen' ('Test rabbits'), *Sudwest Presse*, 1 March 1996.
33. Ortwin Renn, 'Infame Unterstellung' ('Infamous subordination'), *Schwarzwalder Bote*, 10 February 1996.
34. Manfred Bujtor, 'Ausserdem: Sankt Florian' ('Besides St Flovian'), *Sudwest Presse*, 14 May 1996.

4   Risk Management in the United States: The Case of International Paper's Hydro-Dam Re-Licensing Procedure

1. S. Jasanoff, *The Fifth Branch: Science Advisers as Policymakers* (Cambridge, MA: Harvard University Press, 1990).
2. Sheila Jasanoff, *Science at the Bar* (Cambridge, MA: Harvard University Press, 1995); Robert A. Kagan, 'What makes Uncle Sam sue?', *Law and Society Review*, 21 (1988), 734–; Robert A. Kagan, 'Adversarial legalism and American government', *Journal of Policy Analysis and Management*, 10 (1991), 369–406.
3. Robert A. Kagan, 'How much does national styles of law matter?', in R.A. Kagan and L. Axelrad (eds), *Regulatory Encounters: Multinational Corporations and American Adversarial Legalism* (Berkeley, CA: University of California Press, 2000), 3.
4. Sheila Jasanoff, 'American exceptionalism and the political acknowledgement of risk', *Daedalus*, Vol.11 (1991), 61–81.
5. D. Boorstin, *The Americans: The National Experience* (New York: Random House, 1969), 249.
6. Such as the Office of Management and Budget established in 1970, which has been mentioned on numerous occasions in this book.
7. Steven Kelman, *Regulating America, Regulating Sweden: A Comparative Study of Occupational Safety and Health Policy* (Cambridge, MA: MIT Press, 1981).
8. P. MacAvoy, *The regulated industries and the economy* (New York: Norton, 1979); P.J. Quirk, *Industry Influence in Federal Regulatory Agencies* (Princeton, NJ: Princeton University Press, 1981).
9. Christopher Coker, *Twilight of the West* (Boulder, CO: Westview Press, 1998).

10. M. Bernstein, *Regulating Industry by Independent Commission* (Princeton, NJ: Princeton University Press, 1955); G. Kolko, *Railroads and Regulation 1877–1916* (Princeton, NJ: Princeton University Press, 1965); R. Noll, *Reforming Regulation* (Washington, DC: Brookings, 1971).
11. David Vogel, *National Styles of Regulation: Environmental Policy in Great Britain and the United States* (Ithaca, NY: Cornell University Press, 1986), 25–251.
12. Lennart J. Lundqvist, *The Hare and the Tortoise: Clean Air Policies in the United States and Sweden* (Ann Arbor, MI: The University of Michigan Press, 1980).
13. R.V. Percival, 'Checks without balance: executive office oversight of the Environmental Protection Agency', *Law and Contemporary Problems*, 54:4, 127–203.
14. Marc K. Landy, Marc J. Roberts and Stephen R. Thomas, *The Environmental Protection Agency: Asking the Wrong Questions from Nixon to Clinton* (New York: Oxford University Press, 1994); for an insider's discussion regarding this period of environmental regulation see William D. Ruckelshaus, 'Environmental protection: a brief history of the environmental movement in America and the implications abroad', *Environmental Law*, 15 (1985), 455–69.
15. John Quarels, *Cleaning Up America: An Insider's View of the Environmental Protection Agency* (Boston, MA: Houghton Mifflin, 1976), 36. Quote taken from Landy, Roberts and Thomas, *The Environmental Protection Agency*, p. 36.
16. Kelman *Regulating America, Regulating Sweden*.
17. Quirk, *Industry Influence in Federal Regulatory Agencies*; P. Sabatier, 'Social movements and regulatory agencies: toward a more adequate – and less pessimistic – theory of clientele capture', *Policy Sciences*, 6:3 (1975), 301–42.
18. J.Q. Wilson, *The Politics of Regulation* (New York: Basic Books, 1980).
19. Lundqvist, *The Hare and the Tortoise*.
20. C. Coglianese, 'Assessing consensus: the promise and performance of negotiated rulemaking', *Duke Law Journal*, 46 (1997), 1,256.
21. Vogel, *National Styles of Regulation*.
22. Richard B. Stewart, 'The reformation of American administrative law', *Harvard Law Review*, 88 (1975), 1,667–.
23. Jasanoff, *The Fifth Branch*.
24. Lundqvist, *The Hare and the Tortoise*.
25. Landy, Roberts and Thomas, *The Environmental Protection Agency*.
26. Jasanoff, 'American exceptionalism'.
27. Richard H. Pildes and Cass R. Sunstein, 'Reinventing the regulatory state', *The University of Chicago Law Review*, 62 (1995), 1–129.
28. Ibid.
29. Percival, 'Checks without balance'.
30. J.Q. Wilson, *Political Organizations* (New York: Basic Books, 1973).
31. B. Clinton and A. Gore, *Reinventing Environmental Regulation* (Washington, DC: Council on Environmental Quality, 1995).
32. J. Nash and J. Ehrenfeld, 'Codes of environmental management practice: assessing their potential as a tool for change', *Annual Review of Energy and Environment*, 22 (1997), 487–535.

33. S. Georg, 'Regulating the environment: changing from constraint to gentle coercion', *Business Strategy and the Environment*, 3:2 (1994), 11–20.
34. D. Beardsley, T. Davies and R. Hersh, 'Improving environmental management', *Environment*, 39:7 (1997), 6–9, 28–35.
35. C.W. Powers and M.R. Chertow, 'Industrial ecology: Overcoming policy fragmentation', in M.R. Chertow and D.C. Esty (eds), *Thinking Ecologically: The Next Generation of Environmental Policy* (New Haven, CT: Yale University Press, 1997).
36. Coglianese, 'Assessing consensus', pp. 1,255–349.
37. Ibid.
38. J. Freeman, 'Collaborative governance in the administrative state', *UCLA Law Review*, 45:1 (1997), 1–98; J. Freeman and L.I. Langbein, 'Regulatory negotiation and the legitimacy of benefit', *New York University Environmental Law Journal*, 9 (2000), 60–151.
39. L.E. Susskind and J. Secunda, 'The risks and advantages of agency discretion: evidence from EPA's project XL', *UCLA Journal of Environmental Law and Policy*, 17:1 (1999), 67–116.
40. Quotation taken from Coglianese, 'Assessing consensus', p. 1,265.
41. C.H. Koch Jr and B. Martin, 'FTC rulemaking through negotiation', *North Carolina Law Review*, 61 (1983), 275–; H.H. Perritt Jr, 'Administrative alternative dispute resolution: the development of negotiated rulemaking and other processes', *Pepperdine Law Review*, 14 (1987), 863–.
42. Coglianese, 'Assessing consensus', p. 1,261.
43. J.T. Dunlop, 'The negotiations alternative in dispute resolution', *Villenova Law Review*, 29 (1983), 1,429–; W. Funk, 'Bargaining toward the new millennium', *Duke Law Journal*, 46 (1997), 1,351–88.
44. P. Harter, 'Negotiating regulations: a cure for malaise', *Georgetown Law Journal*, 71:1 (1982), 1–118.
45. L. Susskind and G. McMahon, 'The theory and practice of negotiated rulemaking', *Yale Journal of Regulation*, 3 (1995), 133–65.
46. This Act was set to 'sunset' in 1996 but was made permanent by Congress that same year.
47. Coglianese, 'Assessing consensus'.
48. Office of the Vice President, *Accompanying Report of the National Performance Review: Improving Regulatory Systems* (Washington, DC: White House: Office of the President, 1993).
49. Office of the President, Executive Order 12,866; quotation taken from P. Harter, 'Assessing the assessors: the actual performance of negotiated rulemaking', *New York University Environmental Law Journal*, 9:1 (2000), 36.
50. Freeman and Langbein, 'Regulatory negotiation and the legitimacy of benefit'; L.I. Langbein and C.M. Kerwin, 'Regulatory negotiation versus conventional rule making: claims, counterclaims and empirical evidence', *Journal of Public Administration Research and Theory*, 10:3 (2000), 599–632.
51. H. Kunreuther, K. Fitzgerald and T.D. Aarts, 'Siting noxious facilities: a test of the siting credo', *Risk Analysis*, 13 (1993), 301–18; E. Ostrom, 'A behavioral approach to the rational theory of collective action', *American Political Science Review*, 92:1 (1998), 1–22.

52. Environmental Protection Agency (US), *Office of Policy, Planning and Evaluation of the Environmental Protection Agency, Assessment of EPA's Negotiated Rulemaking Activities* (Washington, DC: Environmental Protection Agency, 1987).
53. Coglianese, 'Assessing consensus', p. 1,277.
54. Susskind and Secunda, 'The risks and advantages of agency discretion'.
55. Ibid.
56. Freeman, 'Collaborative governance in the administrative state'; Susskind and Secunda, 'The risks and advantages of agency discretion'.
57. Coglianese, 'Assessing consensus'.
58. Ibid.
59. Ibid., p. 1,322–23.
60. Susskind and Secunda, 'The risks and advantages of agency discretion'.
61. S. Rose-Ackerman, 'American administrative law under siege: is Germany a model?', *Harvard Law Review*, 107 (1994), 1,279–.
62. C. Coglianese, 'The limits of consensus: the environmental protection system in transition: toward a more desirable future', *Environment*, 41:3 (1999), 1–6.
63. E. Siegler, 'Regulatory negotiations and other rulemaking processes: strengths and weaknesses from an industry viewpoint', *Duke Law Journal*, 46 (1997), 1,429–43.
64. Freeman, 'Collaborative governance in the administrative state'.
65. S. Rose-Ackerman, 'Consensus versus incentives: a skeptical look at regulatory negotiation', *Duke Law Journal*, 43 (1994), 1,206–20.
66. C.C. Caldart and N.A. Ashford, 'Negotiation as a means of developing and implementing environmental and occupational health and safety policy', *Harvard Environmental Law Review*, 23 (1999), 141–202.
67. P. Harter, 'Fear of commitment: an affliction of adolescents', *Duke Law Journal*, 46 (1997), 1,389–424.
68. Harter, 'Assessing the assessors', pp. 32–59.
69. C. Coglianese, 'Assessing the advocacy of negotiated rulemaking: a response to Philip Harter', *New York University Environmental Law Journal*, 9:2 (2001), 386–447.
70. Federal Energy Power Act of 1992 (Supplement V, 1994), *Non Federal Power Act, Hydropower Provisions* (Washington, DC: Federal Energy Regulatory Commission).
71. Ibid.
72. It should be noted that throughout Bowater/Great Northern Paper Company's re-licensing procedure dispute with environmental groups the dams remained operational as at the time the original licence had not expired.
73. D.L. Sosland, 'Loser: Bowater/Great Northern Paper Company', *CLF-Conservation Matters* (Spring 1997), p. 27.
74. Harter, 'Negotiated rulemaking'.
75. FERC Order 596, 19 October 1997.
76. M. Bowman, 'Expert testimony before the Subcommittee on Water and Power', Committee on Energy and National Resources, 30 October 1997.

77. 'Electric Power Research Institute, Water Resource Management and Hydro power: Guidebook for Collaboration and Public Involvement'. Report No. TR–104858 (Palo Alto, CA: EPRI, 1995).
78. *Land and Water Associates, Collaborative Processes: More on making the most of FERC's alternative licensing option: Lessons learned from four New England case studies* (Hallowell, Maine: Land and Water Associates, 1999). According to Monte TerHaar, a spokesperson at FERC, approximately 50 per cent of all dam re-licensing procedures use the new regulatory negotiation approach with good results (TerHaar, personal communication, July 2000).
79. Androscoggin is the fourth largest river in Maine. It is 170 miles long and drains 3,450 square miles of Maine and New Hampshire.
80. Kleinschmidt Associates, *Draft Environmental Assessment for Hydropower License* (Pittfield, Maine: Kleinschmidt Associates, 1997).
81. Ibid; Kleinschmidt Associates, *Otis Hydroelectric Company. Otis Hydroelectric Project: Application for New License for Major Project Existing Dam* (Pittfield, Maine: Kleinschmidt Associates, 1997); Kleinschmidt Associates, *International Paper Company. Riley–Jay–Livermore Project: Application for New License for Major Project Existing Dam* (Pittfield, Maine: Kleinschmidt Associates, 1997).
82. Following IP receiving a renewed licence for its four hydropower stations, the town of Jay decided a local code enforcer was no longer necessary.
83. Some environmental NGOs pointed out that the pulp company had direct access to the Governor and hence lobbied him personally to change the environmental laws.
84. For example, R.M. Hunziker, letter to Dan Sosland (on file at International Paper's Androscoggin Mill, 1998).
85. D.L. Sosland and C.A. Blasi, 'Law foundation addresses FERC process', *Bangor Daily News*, 30 December 1996.
86. Harter, 'Negotiating regulations'; P. Harter, 'The political legitimacy and judicial review of consensual rules', *American University Law Review*, 32 (1983), 471–96.
87. It should be noted that at the time of the negotiated rule-making exercise the general public in the area (mainly the city of Jay) had little interest in the exercise. For example, almost no one came to the open meetings that the negotiated rule-making team had scheduled in Jay. There are no clear reasons for this apparent lack of interest among the local population, although one can make a couple of suggestions. First, the instigator of the process, Dan Sosland, an environmental lawyer, was highly trusted by the local community, who had seen him fight against Bowater/Great Northern regarding the dam re-licensing procedure on the Penobscot River. Second, the person within IP who was funding the exercise, Steve Groves, was not an industry insider, but rather an outsider with a background in state regulation. Because of his 'outside' image he too was more trusted by the local community (arguably this was one of the reasons why IP hired him in the first place) than had he been an industry insider.
88. *Bangor Daily News*.
89. On file with IP's Androscoggin Mill, Jay, Maine.

90. Gordon Russell, US Fish and Wildlife Service.

91. A. Giffen, President of Land and Water Associates, telephone interview with author, 11 July 2000; S. Groves, Health Safety and Environment Director of International Paper Company's Jay Mill, interview with author, 22 February 2000; K. Kimball, Research Director at the Appalachian Mountain Club, telephone interview, 12 July 2000; D.L. Sosland, previously senior attorney of the Conservation Law Foundation, telephone interview with the author, 7 March 2000.

92. It was generally believed by the stakeholders involved with the collaborative effort that had state or national regulatory agencies set the criteria for the licences, most of the background studies compiled for the Draft Environmental Assessment would not have been conducted.

93. K. Kimball, Research Director at the Appalachian Mountain Club, telephone interview, 12 July 2000; D.L. Sosland, previously senior attorney of the Conservation Law Foundation, telephone interview with the author, 7 March 2000.

94. Kunreuther, Fitzgerald and Aarts, 'Siting noxious facilities'; Langbein and Kerwin, 'Regulatory negotiation versus conventional rule making'; Ostrom, 'A behavioral approach'.

95. This case was the first example of a collaborative exercise in the country. The process has since been refined and efficiency has increased.

96. One environmental NGO representative that I interviewed argued that in several cases he was involved with the standard command-and-control process came out cheaper than a collaborative approach would have. Or, as Ken Kimball of the Appalachian Mountain Club argued: 'It is like throwing the dice. Sometimes the collaborative approach is cheaper, other times it is more expensive. If there is a massive amount of distrust the collaborative approach will usually be cheaper. Remember, companies want to avoid a Bowater/Great Northern example at all costs' (Interview with Ken Kimball, 11 July 2000).

97. The company itself is a main factor in developing a successful win–win case, but the power of the company should not be overestimated. IP was involved in a re-licensing process in upstate New York involving the same main consultant, Kleinschmidt Associates. Although the dams were eventually re-licensed, it was a long-drawn-out process, with the US Fish and Game Service opposing it.

98. This is the same reason why national environmental groups are not involved in the re-licensing of hydropower stations, passing over these activities to regionally-based NGOs. They simply do not have the time or money to send representatives.

5   Sweden: Barsebäck, Risk Management and Trust

1. C. Ruden, Sven Ove Hansson, M. Johannesson and M. Wingborg, *Att se till eller att titta på-om tillsynen inom miljöområdet (To see or to look at: examination within the environmental area). Ds. 1998: 50* (Stockholm: Fritzes, 1998).

2. T.J. Anton, 'Policy making and political culture in Sweden', *Scandinavian Political Studies*, 4 (1969), 88–102.
3. M. Johannesson, Sven Ove Hansson, C. Ruden and M. Wingborg, 'Risk management – the Swedish way(s)', *Journal of Environmental Management*, 57 (1999), 267–81.
4. Sven Ove Hansson, 'Can we reverse the burden of proof?', *Toxicology Letters*, 90 (1997), 223–8; Sven Ove Hansson, *Setting the Limit. Occupational Health Standards and the Limits of Science* (New York: Oxford University Press, 1998).
5. M. Micheletti, 'Det civila samhallet och staten. Medborgarsammans-lutningarnas roll I svensk politik' ('The civil society and the state: The role of citizen togetherness in swedish politics') (Stockholm: Publica, 1994).
6. S. Kelman, *Regulating America, Regulating Sweden: A Comparative Study of Occupational Safety and Health Policy* (Cambridge, MA: MIT Press, 1981), 118–19.
7. Stepehen Breyer, *Closing the Vicious Circle* (Cambridge, MA: Harvard University Press, 1993); Lennart J. Lundqvist, *The Hare and the Tortoise: Clean Air Policies in the United States and Sweden* (Ann Arbor, MI: The University of Michigan Press, 1980).
8. For example, most of Sweden's engineers, who dominate senior positions in Swedish industry, have been educated at either Chalmers University in Gothenburg or the Royal Technical University in Stockholm, while most of the country's leading economists come from the Stockholm School of Economics.
9. Cited in Lundqvist, *The Hare and the Tortoise*, p. 186.
10. Bo Rothstein, 'Social capital in the Social Democratic State', paper presented at Eleventh International Conference of Europeanists, Baltimore, MD, 29 Feb–1 March 1998.
11. Swedish Statistical Agency, *Välfärd och ojämlikhet I 20-årsperspektiv 1975–1995* (Stockholm: Statistics Sweden, 1997), 327–9.
12. Lundqvist, *The Hare and the Tortoise*, p. 124.
13. Ragnar E. Löfstedt, *Dilemma of Swedish Energy Policy* (Aldershot: Avebury, 1993).
14. Joseph L. Badaracco, Jr, *Loading the Dice: A Five-Country Study of Vinyl Chloride Regulation* (Cambridge, MA: Harvard Business School Press, 1985).
15. Although this is the case in the environmental area, it is different in the occupational health regulatory area. Today 50 per cent of the inspections by the Labour Inspectorate lead to injunctions, which is a higher number than the prosecution numbers by OSHA in the USA. See, for example, W.B. Gray and J.T. Scholz, 'Does the regulatory enforcement work? A panel analysis of OSHA enforcement', *Law and Society Review*, 27 (1993), 177–213; Johannesson *et al.*, 'Risk management – the Swedish way(s)'.
16. Kelman, *Regulating America, Regulating Sweden*.
17. Lennart J. Lundqvist, *Miljövårdsförvaltning och politisk struktur* (*Environmental administration and the political structure*) (Lund: Prisma, 1971), 127.

18. O. Ruin, 'Sweden in the 1970s: Policy making becomes more difficult', in J. Richardson (ed.), *Policy Styles in Western Europe* (London: Allen & Unwin, 1982).

19. Thomas Schelling, *The Strategy of Conflict* (Cambridge, MA: Harvard University Press, 1960).

20. Anders Isaksson, *Den Politiska Adeln: Politikens förvandling från uppdrag till yrke (The political nobility: The changing political administration from commission to profession)* (Stockholm: Wahlstrom & Willstrand, 2002).

21. Badaracco, *Loading the Dice*; kelman, *Regulating America, Regulating Sweden.*

22. See, for example, D.M. Hancock, *Sweden. The Politics of Postindustrial Change* (Hinsdale, II: Holt, Rinehart & Winston, Dryden Press, 1972).

23. Robert C. Sahr, *The Politics of Energy Policy Change in Sweden* (Ann Arbor, MI: The University of Michigan Press, 1985), 16.

24. Lundqvist, *The Hare and the Tortoise.*

25. S. Westerlund, *EG och makten över miljön The EC and the power over environment* (Stockholm: Naturskyddsforeningen, 1992).

26. R.E. Löfstedt, *Risk and regulation: Boat owners' perceptions of recent antifouling legislation* (Risk Management an International Journal, 3: 3 (2001) 33–45; R.E. Lofstedt, *Swedish chemical regulation: An overview and analysis* (Risk Analysis, 23: 2, 411–21; *Swedish Committee on New Guidelines on Chemicals Policy, Non Hazardous Products: Proposals for implementation of new guidelines on chemicals policy SOU 2000: 53*) (Stockholm, Fritzes, 2000).

27. Hansson, 'Can we reverse the burden of proof?'

28. The motion for the bill was taken before the Chernobyl accident.

29. For an overview (in Danish) regarding the history of the Danish view toward the Barsebäck nuclear plant see The Organization for Information about Nuclear Power (OOA), *Hefte Beskrivende den Danske Politiske Holdning til Barsebäck (Booklet describing the Danish political perspective to Barsebäck)* (Copenhagen: OOA, 1993).

30. K.O. Feldt, L. Gyldenkilde, G. Landborn and S. Westerlund, 'Förhindra ett svenskt Tjernobyl!' *(Stop a Swedish Chernobyl), Dagens Nyheter,* 20 October 1992, A-4.

31. *Dagens Nyheter,* 'Danskt smil retar Bildt' ('Danish grin irritates Bildt'), 7 January 1993, A-7.

32. *Dagens Nyheter,* 'Barsebäck måltavla i humorkrig' (Barsebäck is the target for a war of humour'), 11 January 1993, A-5.

33. See, for example, Måns Lönnroth, *Vem Förorenar Sverige? (Who pollutes Sweden?)* (Stockholm: Almäna Förlaget, 1990).

34. S. Bergqvist, *De heta åren (The hot years)* (Malmo: Timbro, 1985).

35. The Swedish Government and Sydkraft, concerned about the Danish reaction to the plant, informed the Danish government about the planning licence six months before the local population.

36. J. Hinderson, 'Barsebäcksverket', *Sydsvenska Dagbladet (The Barsebäck plant),* 22 May 1988. Aircraft landing at Copenhagen's international airport, Kastrup, fly over the Barsebäck plant on the final approach. Since 11 September 2001, there has been a growing concern of the possibility of an aircraft flying into the reactors there.

37. For a discussion on this point see Löfstedt, *Dilemma of Swedish Energy Policy.*

38. For a discussion of Swedish energy policy, please see Jonas Anshelm, *Mellan frälsning och domedag: Om kärnkraftens politiska idehistoria i Sverige 1945–1999 (Between salvation and doomsday: About nuclear power's political idea history in Sweden 1945–1999)* (Stockholm: Brutus Östlings Bokforlag Symposion, 2000); J.M. Jasper, *Nuclear Politics – Energy and the State in the United States, Sweden, and France* (Princeton, NY: Princeton University Press, 1990); Löfstedt, *Dilemma of Swedish Energy Policy*; Sahr, *The Politics of Energy Policy Change in Sweden.*

39. Mikael Gilljam and Sören Holmberg, *Rött blått grönt. En bok om 1988 års riksdagsval (Red, blue, green: A book about 1988 year's national election)* (Stockholm: Swedish Central Statistical Bureau, 1990).

40. The vote on the bill was taken before the Chernobyl accident.

41. Gallup Institute, 'Study on Danish views to the Barsebäck plant', reported in *Berlingske Tidende*, 28 December 1992.

42. G. Steen, *Granskningsrapport angående SKIs handläggning av det s.k.silärendet den 28 July–17 September 1992 (Evaluative study regarding the Swedish nuclear inspectorate dealing with the sieve issue the 28 July–17 September 1992)* (Stockholm: Swedish Nuclear Inspectorate, dnr (classification system) 1.4-921203, 1993).

43. Steen, *Granskningsrapport angående SKIs handläggning.*

44. L. Högberg, 'Milstolpar i svenskt reaktorsamarbete' ('Milestones in Swedish reactor collaboration'), *Nucleus*, 3–4 (1999), 6–15.

45. R.E. Löfstedt, 'Risk communication: The Barsebäck nuclear plant case', *Energy Policy*, 24: 8 (1996) 689–96.

46. Löfstedt, *Dilemma of Swedish Energy Policy*; Löfstedt, 2001.

47. Of course there are always exceptions. Once the Barsebäck stations were reopened not all policy-makers were happy with this outcome. Most notably the Environmental Minister, Olof Johansson (then head of the Centre Party), announced that the Swedes would drive their nuclear reactors until they exploded. See Löfstedt, 'Risk communication'.

48. Steen, *Granskningsrapport angående SKIs handläggning.*

49. M. Uhrwing, *Intressepresentation i brytningstid. En studie av intressepresentation i några miljöpolitiska organ (Interest representation in times of breakage. A study of interest representation in a few political environmental institutions)* (Gothenburg: Department of Political Science, University of Gothenburg, 1995).

6   Risk Management in the UK: The Case of Brent Spar

1. House of Lords, Select Committee on Science and Technology, *Science and Society* (London: The Stationary Office, 2000).

2. From J.E.S. Hayward, 'National aptitudes for planning in Britain, France and Italy', *Government and Opposition*, 9:4 (1974), 397–410, reprinted in G. Jordan and J. Richardson, 'The British policy style or the logic of

negotiation?', in J. Richardson (ed.), *Policy Studies in Western Europe* (London: Allen & Unwin, 1982), 81.

3. R. Macrory, 'The United Kingdom', in G. Enyedi, J. Giswijt and B. Rhode (eds), *Environmental Policies in East and West*, (London: Taylor & Francis, 1997), 87, quotation taken from John McCormick, 'Environmental policy in Britain', in U. Desai (ed.), *Environmental Politics and Policy in Industrialised Countries* (Cambridge, MA: MIT Press, 2002), 124.

4. It should be noted, however, that the use of cost-benefit analysis in helping to determine the costs of regulation is more pronounced in the UK than in many other European nations.

5. See, for example, House of Lords, *Science and Society*; RCEP, *Setting Environmental Standards* (London: The Stationary Office, 2000); UK Cabinet Office, Strategy Unit Report, *Risk: Improving Government's Capability to Handle Risk and Uncertainty* (London: Strategy Unit, The Cabinet Office, 2002).

6. For a great review on the history of UK environmental regulation please see David Vogel, *National Styles of Regulation: Environmental Policy in Great Britain and the United States* (Ithaca, NY: Cornell University Press, 1986).

7. E. Ashby and M. Anderson, *The Politics of Clean Air* (Oxford, Clarendon Press, 1981); Lord Asquith 1949, *Edwards* v. *National Coal Board* 1KB;1949. 1AII ER 743, p. 712 and p. 747, a case interpretation of S. 102 (8) of the Coal Mines Act 1911.

8. J. McCormick, *British Politics and the Environment* (London: Earthscan, 1991).

9. A.E. Dingle, 'The monster nuisance of all: landowners, Alkali manufacturers and air pollution 1858–1862', *Economic History Review*, 35 (1982), 529–48.

10. Speech presented at the 88th Environmental Health Congress, Harrogate, 30 September 1980, p. 900; quotation taken from Vogel, *National Styles of Regulation*, p. 86.

11. Martin Weiner, *English Culture and the Decline of the Industrial Spirit, 1850–1980* (Cambridge: Cambridge University Press, 1981).

12. Ashby and Anderson, *The Politics of Clean Air*.

13. McCormick, 'Environmental policy in Britain'.

14. David Storey, 'An economic appraisal of the legal and administrative aspects of water pollution control in England and Wales, 1970–1974', in T. O'Riordan and Ralph C. D'Arge (eds), *Progress in Resource Management*, Vol. 1 (New York: Wiley, 1979), p. 263; quotation taken from Vogel, *National Styles of Regulation*, p. 89.

15. J. Hayward and R. Berki, *State and Society in Contemporary Europe* (Oxford: Robertson, 1979).

16. Timothy O'Riordan and Brian Wynne, 'Regulating environmental risks: a comparative perspective', in Paul Kleindorfer and Howard Kunreuther (eds), *Insuring and Managing Hazardous Risks: From Seveso to Bhopal and Beyond* (Berlin: Springer Verlag, 1987).

17. For a recent example see Holly Welles and Kirsten Engel, 'Siting solid waste fills: the permit process of California, Pennsylvania, the United Kingdom, and the Netherlands', in Robert A. Kagan and Lee Axelrad (eds), *Regulatory Encounters: Multinational Corporations and American Adversarial Legalism* (Berkeley, CA: University of California Press, 2000).

18. Sonja Boehmer-Christiansen and Jim Skea, *Acid Politics* (London: Belhaven Press, 1991); Sheila Jasanoff, *Risk Management and Political Culture* (New York: Russell Sage Foundation, 1986); Sheila Jasanoff, 'Cultural aspects of risk assessment in Britain and the United States', in Branden B. Johnson and Vincent T. Covello (eds), *The Social Construction of Risk* (Leiden: D. Reidel, 1987), 359–97.

19. W.G. Carson, *The Other Price of Britain's Oil: Safety and Control in the North Sea* (New Brunswick, NJ: Rutgers University Press, 1982).

20. Brian Wynne, *The Hazardous Management of Risk – Comparative Institutional Perspectives* (Berlin: Springer Verlag, 1986).

21. House of Lords, *Science and Society*.

22. Survey found in House of Lords, *Science in Society*, p. 88.

23. Anthony Giddens, *The Consequences of Modernity* (Cambridge: Polity, 1990); Barbara A. Misztal, *Trust in Modern Societies* (Cambridge: Polity, 1996), Susan J. Pharr and Robert D. Putnam, *Disaffected Democracies: What's Troubling the Trilateral Countries?* (Princeton, NJ: Princeton University Press, 2000); Robert D. Putnam, *Bowling Alone: The Collapse and Revival of American Community* (New York: Simon & Schuster, 2000).

24. Ragnar Löfstedt, *Risk Evaluation in the United Kingdom: Legal Requirements, Conceptual Foundations and Practical Experiences with a Special Emphasis on Energy Systems* (Stuttgart: Centre for Technology Assessment, 1997); Philip Lowe and Stephen Ward, *British Environmental Policy and Europe: Politics and Policy in Transition* (London: Routledge, 1998).

25. RCEP, *Setting Environmental Standards*.

26. House of Lords, *Science and Society*; House of Lords, Select Committee on Science and Technology, *Science and Society: Evidence* (London: The Stationery Office, 2000).

27. Lord Woolf (Lord Chief Justice), 'The Professor David Hall Lecture: Environmental risk: responsibility of the law and science', 24 May 2001, School of Oriental and African Studies, London.

28. Cabinet Office, *Open Government* (London: HMSO, 1993).

29. R. Macrory, 'Environmental Law: shifting discretions and the new formalism', in O. Lomas (ed.), *Frontiers of Environmental Law* (London: Chancery Law, 1991).

30. H. Buller, 'Reflections across the channel: Britain, France and the Europeanization of national environmental policy', in P. Low and S. Ward (eds), *British Environmental Policy and Europe* (London: Routledge, 1998).

31. Lord Woolf, 'The Professor David Hall Lecture'.

32. S. Jasanoff, 'Civilization and madness: the great BSE scare of 1996', *Public Understanding of Science*, 6 (1997), 221–32.

33. NERC, *Scientific Group on Decommissioning Offshore Structures First Report* (Swindon: NERC, 1996); NERC, *Scientific Group on Decommissioning Offshore Structures Second Report* (Swindon: NERC 1998).

34. Rudall Blanchard Associates, *Brent Spar Abandonment BPEO, prepared for Shell U.K. Exploration and Production* (London: Shell, 1994); Rudall Blanchard Associates, *Brent Spar Abandonment Impact Hypothesis, prepared for Shell U.K. Exploration and Production Limited* (London: Shell, 1994).

35. Eggar, quoted in C. Rose, *The Turning of the Spar* (London: Greenpeace, 1998). Quotations taken from G. Jordan, 'Indirect causes and effects in policy change: the Brent Spar case', *Public Administration*, 76 (1998), 717.
36. Jordan, 'Indirect causes and effects in policy change', p. 713–40.
37. Of Greenpeace's £1.4 million budget for the campaign, £350,000 was spent on the media.
38. Rose, *The Turning of the Spar*.
39. Greenpeace, 'Memorandum by Greenpeace', in House of Lords, *Science and Society: Evidence*.
40. Quotation taken from C. Clower, G. Jobes, A. Cramb and D. Milward, 'Shell "caves in" over dumping of Brent Spar', *Daily Telegraph*, 21 June 1995, p. 1.
41. Quotation taken from Rose, *The Turning of the Spar*.
42. Clower, *et al.* 'Shell "caves in"'.
43. Rose, *The Turning of the Spar*.
44. NERC, 1996 *Scientific Group on Decommissioning Offshore Structures*, NERC, 1998 *Scientific Group on Decommissioning Offshore Structures*.
45. NERC, *First Report*; NERC, *Second Report*.
46. Ibid.
47. Derek Osborn, 'Some reflections on UK environmental policy, 1970–1995', *Journal of Environmental Law*, 9 (1997), 10.
48. R. Gribben, 'Shell wins permission to sink redundant oil rig in Atlantic', *Daily Telegraph*, 17 February 1995.
49. Before becoming the UK Energy Minister, Tim Eggar was on the Board of Charterhouse Petroleum. After leaving the Government he became Chairman of AGIP UK, Monument Oil, and the offshore contractors, Kellog.
50. See Rose, *The Turning of the Spar*.
51. Ibid.
52. Ibid.
53. De Ramsey 1995, cited in Ragnar E. Löfstedt and Tom Horlick-Joness, 'Environmental regulation in the UK: politics, institutional change and public trust', in George Cvetkovich and Ragnar E. Löfstedt (eds), *Social Trust and the Management of Risk* (London: Earthscan, 1999), 83.
54. *The Times*, 'Grow up, Greenpeace: A little more responsibility is now required', 6 September 1995.
55. Rose, *The Turning of the Spar*; T. Rice and P. Owen, *Decommissioning the Brent Spar* (London: Routledge, 1999).
56. Rose, *The Turning of the Spar*.
57. OLR, 1996.
58. MORI, 1995.
59. Vogel, *National Styles of Regulation*.
60. Jasanoff, 'Civilization and madness'.
61. Greenpeace, 'Memorandum by Greenpeace'.
62. It should be noted that Greenpeace and other environmental organizations have tried to discredit science, arguing that a broad range of deliberation is needed, so they too can participate in the risk management process J.S. Gray, 'Statistics and the precautionary principle'. Marine Pollution Bulletin (1990) 21: 174–6; J.S. Gray and J. Brewers, 'Towards scientific

definition of the precautionary Principle', Marine Pollution Bulletin 1996 26: 768–71; J.S. Gray, D. Calamari, R. Duce, J.E. Portmann, P.G.Wells and H.L. Windom, 'Scientifically based strategies for marine environmental protection and management', Marine Pollution Bulletin 1991, 22: 432–40.

63. Vogel, *National Styles of Regulation*.
64. Shell did introduce a dialogue approach once it had decided that it would not dump the oil storage buoy in the deep sea.
65. Rose, *The Turning of the Spar*.
66. For a discussion regarding the policy vacuum and risk communication please see Douglas Powell and William Leiss, *Mad Cows and Mothers Milk: The Perils of Poor Risk Communication* (Montreal: McGill-Queen's University Press, 1997).
67. Rose, *The Turning of the Spar*.

7   Conclusions: Integrating Trust into Risk Management

1. R. Löfstedt, (1996) *Risk Communication*.
2. NRC, *Improving Risk Communication* (Washington, DC: National Academy Press, 1989).
3. Jacques Thomassen, 'Support for democratic values', in H.-D. Klingman and D. Fuchs (eds), *Citizens and the State* (Oxford: Oxford University Press, 1995).
4. Ragnar E. Löfstedt and Tom Horlick-Jones, 'Environmental regulation in the UK: politics, institutional change and public trust', in George Cvetkovich and Ragnar E. Löfstedt (eds), *Social Trust and the Management of Risk* (London: Earthscan, 1999), 73–88.

Index